Foraging New Zealand

Over 250 plants and fungi to forage in Aotearoa

Peter Langlands

PENGUIN BOOKS

Kiokio

I dedicate this book to my parents
Elizabeth and Ian Langlands,
who made the brave journey
across the world to Aotearoa.

For me, this is a place where two
worlds unite and I am forever
grateful for my life here.

Contents

Kānuka in flower — a sign of summer.

Foraging, a personal journey

For me, foraging is the ultimate integration of my outdoor interests: bird research and watching, recreational fishing, and photography. Spending time in a wide range of outdoor environments set the stage for my interest in foraging, but I, like many people, have foraged for years without realising it, even before the term 'foraging' in the human context became part of the vernacular. When I was growing up, foraging was a term that was not used very often and even then only in reference to wildlife, not people. All that changed around 20 years ago when a trend emerged to supply locally gathered foods to top-level restaurants. The ethos of focusing on locally available food and using it as a unique selling point for creative dishes became popular. People such as Hugh Fearnley-Whittingstall and René Redzepi set the stage in Europe, and it arrived in New Zealand with chefs such as Vaughan Mabee focusing on sourcing food for his restaurant from the local region.

Meanwhile, outside the commercial kitchen, people were increasingly discovering a sense of satisfaction from spending time in nature and gathering their own food from their back doorstep. This was tasty food, which was not commercially available, and it offered novel flavours that were often more intense than cultivated produce. The arrival of Covid-19 restrictions further extended my interest, and the interest of others, in getting to know our own home ground well. Many more people discovered that foraging gives you a reward for spending time in nature. And so, within a couple of decades, the term foraging has arisen from the background to become a thing that humans do.

Looking back, I realise I was already foraging when I was a small child, picking wild blackberries and gooseberries with my mother. I must have been about 10 when I started nibbling on samphire while watching rare wading birds at Te Waihora (Lake Ellesmere). I was not shy about putting nature in my mouth! As an adult, I have always spent much of the year outdoors, so further foraging crept in naturally. More recently, the joy of foraging has blossomed into a profession. Each week, my partner, Jackie, and I run foraging workshops, tours and webinars, sharing our knowledge of wild foods and their utilisation. Each week we find new species, in new places, with new people, creating many unique experiences. It is that exciting sense of discovery that keeps foraging interesting. You always come across the unexpected, and the spontaneity of nature is of course ever changing. Sharing this knowledge of using wild foods is a rewarding and learning experience.

Foraging keeps my sense of curiosity alive; not only is there a never-ending diversity of species to discover — along with new flavours and many health benefits — but also there are numerous ways to prepare wild foods. The internet and social media have certainly opened up opportunities for using foraged foods, from tapping trees for sap, to collecting wild grains and sprouting them, to using pinecones for making a flavoured syrup. The options are mind-boggling, and collectively there are more ideas shared than ever: picking just one to work on is the challenge! Tasting the foods as they are picked is a fun part of the foraging experience. Even more rewarding is sharing what you have gathered with others. I have found that a picnic of foraged foods sets up lifelong memories: it is both a sensory and a social engagement. In many ways the once-lost art of foraging and the lost art of picnics merge together seamlessly.

Yet, despite all the opportunities, it is important to be careful with foraging. Knowing what to safely forage is the key, and that is the primary aim of this field guide. Taking the time to make an accurate identification is crucial because, as we regularly say, *There is not much room for error.* This is especially the case when it comes to fungi, but also applies to an array of wild plants. You have to be cautious; and if eating a new wild plant, it pays to try only a little bit initially as each person will have different reactions to different types of food. When you decide to go out gathering, take it one step at a time, focusing on one edible species, building up your knowledge and identification techniques as you go, and the results of your developing skill base will be tangible.

The way forward for foraging is exciting as knowledge is shared. It is my personal hope that, through foraging, people will become not only aware of but also advocates for maintaining a healthy local environment. Foraging slows you down and makes you conscious of your surroundings, and there is no doubt that the more relaxed you are the more you will see. Rather than being perplexed by the diversity before you, step back and look around and assess the environment to ensure that the soil and air are clean. Take a moment to get to know the place.

In many ways, foraging will take you off the beaten track. I love turning off country roads that I have not previously travelled down and coming across new foraging prospects. These ever-changing opportunities to bring home incredible bounty to the table make me feel a sense of rebellion. I know my food and where it comes from; there are no shops or other intermediaries between it and me. Foraging is a personal journey that keeps me motivated to get outside — so welcome to the journey!

Peter Langlands
Wild Capture Foraging
April 2024

Black tree ferns growing under a pine forest — many ferns are edible.

Introduction

In Aotearoa New Zealand we are lucky to have a unique combination of our native (and endemic) species as well as a vast array of plants introduced from the rest of the world. That gives us over 7500 edible species to consider, so having to focus on the 250 main species for this book was certainly a challenging — though rewarding — experience. Those included in this book are selected for their flavour, being generally recognised as safe to eat, and for being readily available. This last point means that endangered species are not included. I have covered only plant-based foraging — edible creatures would fill another book, and foraging for healing, cosmetics, cleansers, perfumes or dyeing already appears in many other books in the market. Despite these selection criteria, there are still many plants not included because there wasn't room; after all, no one wants to take a heavy brick of a book with them into the field.

You'll notice that each plant has at least one common name, often more, plus a scientific name, sometimes a former scientific name; and for the plants that have been in this country a long time a Māori name too. These multiple names make identification a little more challenging, which is only exacerbated because both Māori and English names can apply to several quite different species. As if that wasn't confusing enough, some plants have been reclassified, and so shifted into a completely different family. And those family names can also change. If your head isn't already in a whirl, also be aware that names used by those from, say, America might be different from those from the United Kingdom, or India. Foragers need nimble minds to keep track of all this and must allow for the fact that some of the designations given here will inevitably alter in the future or won't be what you currently use. In fact, foragers need to keep on top of research for lots of aspects, not least toxicity.

Toxicity

Despite being consumed sometimes for millennia, many commonly foraged foods have been insufficiently researched, so foragers must deal with a lot of knowledge gaps. What is recommended in these pages today will not necessarily be the case in the future when further tests are done. Several ferns and fungi, for instance, were deemed safe 10 years ago but have now been associated with toxic accumulation. Also, just because a species has medicinal properties doesn't mean it is safe for everyday consumption.

Meanwhile, other species may be perfectly safe to eat in themselves but make sure to give your foraged foods a decent wash, especially if you suspect they have been sprayed by herbicides or dogs, or grow close to such things as algae bloom, old landfills, dioxin runoffs from timber mills, or a watercourse containing high levels of nitrates or agrochemicals. That is before you have ensured that you are not confusing your foraged item with a similar poisonous variety. Thus the message with foraging is to proceed with caution: if in doubt, do not pick.

There is only limited space in this book for descriptions, so if unsure of identification double-check with reliable sites, such as iNaturalist.NZ and the New Zealand Plant Conservation Network. Be wary of relying too heavily on social media — if you look at identification requests there, you'll see differing opinions from people who might be experts but more often are amateurs guessing, often very haphazardly.

While foraged items are generally full of vitamins, minerals and beneficial chemicals, some considered edible will still have low levels of toxicity. This is natural, as plants often contain toxins to ward off predators. Many of our everyday foods contains toxins, from green potatoes to raw kidney beans to bitter zucchini to apple seeds. Sometimes toxins are removed by cooking or peeling — but this is not always the case. Some toxins will do no harm if eaten in moderation, but they could have negative side effects for those who are pregnant or feeding an infant, for the very young or for the sick. Some people develop unexpected allergies, so even if you know that a new find is not poisonous, always taste small amounts first.

A number of plants in this book, such as wood sorrels and *Rumex* leaves, contain oxalic acid, or oxalates, a naturally occurring toxin. Spinach, cereal grains, raspberries and dates, among many everyday foods, also contain oxalic acid, which in addition is produced by our own bodies. So, oxalic acid is generally not an issue and will usually pass right through our systems, and the many positive nutrients in these plants mean they are important to our diets.

Oxalic acid binds to any calcium we have consumed, which means there is less of this valuable nutrient available for us to absorb, potentially leading to calcium deficiency if eaten too regularly. Furthermore, excess acid could form crystals in our system, which might lead to kidney stones, especially in those with kidney disease or who have existing gut problems. We tend to limit our intake of oxalic acid unconsciously simply by balancing the often strongly flavoured food with other ingredients, but cooking it — especially by boiling and discarding the water — or fermenting it will further reduce the levels consumed. The good news is that Oxalic acid is also broken down in our gut by *Oxalobacter formigenes*. This presence of this bacterium can increase by more than 100 per cent when you start to eat wild foods. This is all to say as well, that its incredibly unlikely you would ever eat an amount of foraged food containing oxalic acid for it to actually do any harm.

It should be noted that I have not included the Boraginaceae family, although it includes plants that have commonly been regarded as medicinal herbs — such as comfrey (*Symphytum* spp.) — or attractive garnishes, as with the flowers of green alkanet (*Pentaglottis sempervirens*), or as a useful addition to traditional recipes; for instance, in Europe the leaves of borage (*Borago officinalis*) have long been used in soup, sauces, pasta fillings, and herbal tea and gin. The leaves of many of these plants have a pleasant cucumber flavour, making them particularly delicious added to drinks, not to mention the visual appeal of an ice cube containing a borage flower. And the forager is likely to find members of the family across much of the country: take forget-me-not (*Myosotis sylvatica*) as a particularly well-known example. Their often prickly stems might cause irritation, as with viper's bugloss (*Echium vulgare*), but their leaves have been a welcome addition to salads and all sorts of meals. However, recent research has concluded that these plants contain potentially harmful pyrrolizidine alkaloids. The level of these alkaloids seems to vary from plant to plant, and indeed may be minimal in many instances, particularly in the case of forget-me-knot, though those flowers are so small they are barely worth gathering. So, small quantities of these plants may do no harm, but large amounts or consumption over an extended period of time could lead to serious health issues. If you nibble them, do so at your own risk.

There are a number of plants in our everyday diet that have a highly edible part alongside a toxic part. For example, rhubarb (*Rheum* x *rhabarbarum*) has toxic leaves, and taro (*Colocasia esculenta*) must be prepared correctly to be consumed. The same caution is needed with many wild species worth foraging. I have added specific details about food safety in the individual sections; please read them carefully as some species are more hazardous than others. One of the biggest risks, of course, is picking a deadly specimen by mistake, which is why identification is so vital.

Identification

To make your identification, observe different aspects such as the overall form of the plant, the leaf shape, colour and undersides, the presence or absence of hairs or thorns, the shape and colour of flowers, the smell, and, if you are looking at fungus, the spore print and presence or otherwise of gills. Do not rely on a quick recognition of just one element, as similar species abound. And even if you have identified correctly, make sure there are no other species sneakily mixed in.

Take note of the environment you are in as this will give you clues about what you will find. Plants thrive in different conditions, so altitude, closeness to the coast, the soil, the amount of sunlight or moisture, and the companion species are all important factors in determining what will be growing nearby.

Introduced species have spread from gardens, so urban wastelands or abandoned settlements are good places to search. Likewise, people transport plants wherever they travel — so look alongside roads, rivers and railway lines. These are 'edge environments', a phrase that crops up a lot in this book and refers to the margins between ecosystems or types of landscape, such as between farm and open countryside, or on the outskirts of settlements, or the area around forests or between bush and pasture.

Distribution

The spread of plants is another ever-changing topic to keep foragers on our toes, so what is given here will also change over time. Many of the plants recommended are invasive — so foraging seeds in particular is a good thing, but even so their rampage across the country is likely to increase. And climate change will mean some plants will reach new areas, while others might retreat. As a result, treat the distribution details here as rough guidelines only.

Gorse on a hillside on Banks Peninsula, Canterbury.

A diversity of trees grow along the edge of Lake Wakatipu, Otago.

These days, there are more restrictions on what can be brought into Aotearoa than used to be the case, but despite this some future newcomers will inevitably find New Zealand conditions very much to their liking and jump the garden fence. Others will hitch rides on boat bottoms or in imported products. This means you need to be vigilant with identification: these intruders might be poisonous and might be confused with edible species. Of course, they might also add to the list of wonderful items to forage.

Unless otherwise mentioned, the native species included in this book are not threatened — at least not at the time of writing. Again, climate change, or increasing competition from spreading weeds, introduced animals or human interaction, may alter their status, so keep alert in case your favourite bit of browsing has just been put on the at-risk list. Regardless of that, whenever you harvest native species do not destroy the entire plant, do not remove every specimen in an area, and do not trample other plants nearby. If you are not collecting the seeds, do not remove them but scatter them instead. *Under*pick rather than *over*pick.

Permits

There are some places — such as national parks, areas of scenic or conservation value, or reserves used for special purposes, be they research or plant or wildlife protection — where it is illegal to forage without a collecting permit. The Department of Conservation would be your first port of call to secure one of these. You must also seek permission from the appropriate authority to forage in public parks, forest plantations, council plantings, land with covenants on it and obviously from private property. Māori land is private, and permission must be sought to enter it, and correct tikanga must be observed. Note that there might be temporary local bans in specific areas because of a rāhui or to allow certain ecosystems to regenerate or to avoid spreading kauri dieback.

Some species are regarded as a taonga by Māori, so respect local protocols, and note that native plants in particular should not be collected in certain places. As the legislation states: 'every person commits an offence who takes any protected native plant that is growing on any Crown land, or in any State forest land or public reserve, or on any road or street, or who, without the consent of the owner or occupier of any private land, takes any protected native plant that is growing thereon'.

Overall, recreational (though not commercial) harvesting of wild seaweeds is allowed along most of our coastline but not from marine reserves. Harvesting of seaweeds is controlled by the Fisheries Act, and care needs to be taken to check any local regulations, especially within mahinga mātaitai or taiāpure areas. Although there are some private beaches and properties that border the sea, everyone is allowed up to the high-tide mark; access, however, might be restricted if you need to cross private land to get there.

Amaranthaceae

The striking red *Amaranthus* flower.

17

Cereals, such as wheat or rye, come from grasses; however, other starchy grains that are not produced by grasses can be cooked and consumed in the same way. These gluten-free seeds are known as pseudocereals and include many members of the Amaranthaceae family (see also *Rumex*), which often have edible leaves as well as seeds; quinoa is a notable member of this family and has become widely used for its grain-like seeds.

The *Amaranthus* species were initially cultivated in Aotearoa New Zealand for their decorative value, but having started to spread to the wild they are now a target for foragers because of their highly nutritious seeds as well as edible leaves. And with some members of the wider Amaranthaceae family, it is the leaves that are the main focus: for instance, silver beet. Its large, crinkly, dark-green leaves with their distinct mid-rib are recognisable from the garden, but keep an eye out as the plant has become naturalised along many coastal margins, often just above the high-tide line, enjoying slightly saline soils. This fondness for salty environments is shared by a number of the species.

Within the same family are the *Atriplex* species, commonly known as orache or saltbush, which have been eaten as a cereal for millennia, though the species in New Zealand are usually harvested for their leaves. While the spear-leaved orache has become a bit of a weed across the country, there are several other members of this genus that only appear in a few places so are best left unharvested. This includes some endemic varieties, such as grey saltbush (*Atriplex cinerea*), which have different leaves and are semi-fleshy and only patchily spread — several only appear on the Chathams. Even common orache, *Atriplex patula*, is not actually widespread in Aotearoa New Zealand.

Also included here are some other plants in the Amaranthaceae family that are not notable for their seeds but are still worth foraging for their leaves or stems.

Amaranthus species

SCIENTIFIC NAMES

Amaranthus albus, Amaranthus blitoides, Amaranthus caudatus, Amaranthus cruentus, Amaranthus deflexus, Amaranthus graecizans, Amaranthus hybridus, Amaranthus hypochondriacus, Amaranthus lividus, Amaranthus powellii, Amaranthus retroflexus, Amaranthus spinosus, Amaranthus tricolor, Amaranthus viridis

OTHER NAMES (DEPENDING ON VARIETY)

Foxtail amaranth, green amaranth, love-lies-bleeding, mat amaranth, Powell's amaranth, prostrate pigweed, purple amaranth, redroot, tumbleweed amaranth

DESCRIPTION

A popular decorative species that, being self-fertile, fast-growing and a producer of copious seeds, has spread over the garden fence. Although usually smaller, some varieties grow 1-2m high, while *Amaranthus deflexus* and *Amaranthus lividus* sprawl across the ground. *Amaranthus* have well-spaced, narrow-tipped, oval leaves with distinct parallel veins (a bit like beech leaves), and while each variety has different features, they are often splashed with bright red, pink, purple or gold, be it in leaf markings, or their stems, or their often flamboyant spires or tassels of flowers, or their seedheads. Note, though, that *Amaranthus powellii* is green-leaved and green-flowered.

DISTRIBUTION

Amaranthus powellii is probably the most widely spread *Amaranthus* weed in Aotearoa New Zealand, but *Amaranthus retroflexus* is not far behind and most varieties can be found in urban areas. *Amaranthus deflexus* generally grows between Canterbury and Waikato, *Amaranthus lividus* mostly in the North Island, *Amaranthus viridis* only in the top half, while other species have a patchier spread. *Amaranthus tricolor* is easily sourced from garden centres.

HABITAT

Often found infiltrating crops and orchards in dry, sunny sites or in urban wasteland areas.

SEASONALITY

Harvested through summer and autumn, they die off in winter.

FORAGING TIPS

Don't harvest the older leaves as they become slightly bitter and tougher. The seeds are ready to harvest when they come off the plant with gentle rubbing, or else cut off the dying flowers and place in bags to dry, where the seeds will collect after a good shake.

CAUTIONARY NOTES

Do not harvest where the ground might be high in fertilisers or metals or affected by farm or industrial runoff as the plants readily absorb nitrates. The leaves contain oxalic acid (see page 12), so use sparingly.

Key edible uses

Amaranthus tricolor produces fewer seeds so is primarily used for its edible stems and leaves, but most other *Amaranthus* species have huge plumes of flowers that produce masses of protein-rich seeds. The plants can be eaten as microgreens or the young, mild-tasting leaves can be added raw to smoothies or salads (adding a flash of colour if you find a red-leaved version); or cooked, along with the chopped stems, and used as a spinach substitute. Eaten by the Aztecs, the nutty seeds can be eaten dried or fresh, popped like popcorn or boiled with water into a porridge. Alternatively, they can be used like rice, or ground into a flour for baking or as a thickener. The plant is high in vitamin C, protein, amino acids, antioxidants, iron, manganese and various beneficial minerals.

Spear-leaved orache

SCIENTIFIC NAME
Atriplex prostrata

OTHER NAMES
Fat hen, hastate orache, thin-leaf orache, triangle orache

DESCRIPTION
A fresh green, soft-leaved plant growing to 0.7m high with small green spires dotted with pink which produce clusters of comparatively large seeds growing from central stems and tiny lance-like leaves. The fresh green leaves on the lower stems are shaped like arrowheads with serrated edges.

DISTRIBUTION
Found throughout drier parts of Aotearoa New Zealand.

HABITAT
Found in well-drained soils along edge environments, often colonising areas where the land has been cleared, for example for suburban development.

SEASONALITY
An annual green with leaves available through spring, summer and autumn; the seeds harvested in autumn.

FORAGING TIPS
Often grows in dense beds and can easily be harvested in large bunches.

CAUTIONARY NOTES
Check soil quality at the harvesting site as the leaves will absorb nitrates from runoff. When foraging from dense beds, check that there are no imposter plants, such as nightshades, growing in among the stems.

Key edible uses

The fresh-growth stems and leaves can be juiced raw or put into green smoothies, or when cooked they are a useful spinach substitute, served as a side green or added to stir-fries or pasta dishes. The chopped older leaves are suited to fermentation in either sauerkraut or kimchi. While not prolific, the seeds can be used as a grain.

Opposite The lance-like leaves gives the spear-leaved orache its name.

Samphire

SCIENTIFIC NAME
Salicornia quinqueflora

MĀORI NAME
Ureure

OTHER NAMES
Sarcocornia quinqueflora, beaded samphire, chicken claws, glasswort

DESCRIPTION
A succulent with cylindrical, jointed spears up to 10cm high and less than 0.5cm wide. Growing in dense mats, its fresh growth is bright light-green, fading to pinkish-brown and grey, dying away to woody basal mats over winter. The tiny yellow flowers appear from bumps on the top of the stalk.

HABITAT
Saline flat environments primarily just above the high-tide line, and also in patches of rocky shoreline just above the high-tide mark up to areas where salt spray penetrates. Often common in areas that regularly get inundated with saline water or in brackish water.

SEASONALITY
An annual species with fresh growth from September to early December, reaching peak growth from December to January and then flowering in late summer before drying away to basal woody mats over winter. Best harvested from late December until early March with the fresh green growth.

FORAGING TIPS
Use a sharp knife to forage the top half of the spear, leaving the basal root intact. Wash well as pathogens can occur if the water is polluted.

DISTRIBUTION
Found throughout Aotearoa New Zealand around estuarine environments in river margins, estuaries, habours and coastal lagoons.

Key edible uses

With a crunchy texture and a salty asparagus-like flavour, it is high in vitamins and can be eaten raw as a garnish or added to a salad or chopped into mayonnaise in place of pickle. Alternatively, to reduce its saltiness it can be lightly blanched and used either as a side green or dropped into soups and stews to provide a salty zest. It is ideal for pickling in vinegar.

Sea blite

SCIENTIFIC NAME
Suaeda novae-zelandiae

OTHER NAME
Seepweed

DESCRIPTION
A non-threatened Aotearoa New Zealand succulent that creeps on the ground or in small mats, producing multiple branches of woody stalks holding clusters of small, succulent, oval leaves in tones of yellow, green, brown, red and purple, with tiny white flowers on the stalks at the base of the leaves.

DISTRIBUTION
Found in coastal locations in patches throughout Aotearoa but with a stronghold in the Auckland region.

HABITAT
Salt-marsh environments, liking alkaline soils. Often grows in mats with other species such as samphire. Often found growing on flat salt-marshes or in among rocks.

SEASONALITY
Available year-round.

FORAGING TIPS
Only harvest the tips, letting the plant continue to grow, but check the water quality first and also wash thoroughly.

Inset The small, white flowers of the sea blite are visible against the green tones of this coastal succulent.

Key edible uses

With its pleasant tangy, salty flavour, the fresh growth of this often-overlooked succulent can be eaten raw or lightly blanched to reduce the saltiness, used as a garnish, put into salads, added to soups, made into a relish or pickled in vinegar. The plant is a good source of protein, a number of vitamins and beneficial compounds.

Apiaceae

Wild carrot flower in bloom.

25

Although this family is now named Apiaceae, the earlier label of Umbelliferae, or more commonly umbellifers, is a useful key to identification as that name comes from the umbrella-shaped flower clusters, known as umbels. The flower clusters forming each umbel appear on an array of short stems coming from a single point at the top of the stalk, like the ribs of an umbrella. This family includes herbs and vegetables, such as fennel, coriander, parsley, celery, carrot and parsnip, so it is no surprise that they make good subjects for foraging. In fact, many of these garden varieties have gone wild, so as well as the species in the following pages, don't be surprised if you encounter such common garden favourites as Italian parsley. But also note, the young foliage could be confused with that of hemlock (see page 46). It is critical to be aware that the Apiaceae family includes some highly poisonous species, so careful identification is essential. If in doubt, do not pick.

Distinctive flavour is a common feature of many of these plants. The taste of aniseed, for instance, is particularly present in a number of species in the following pages, as well as in another not included here but worth noting: the little native plant, *Anisotome aromatica*, also known as kopoti or pinakitere. Although it is fairly common in upland areas from the Central Plateau south, it isn't often foraged; perhaps because of its small size or because you have to trek to find it. A few leaves would, though, spice up a tramper's usual tucker. There are other native Apiaceae that are edible, but many are endangered so should be left alone. In addition, the *Aciphylla* speargrasses, such as golden Spaniard, *Aciphylla aurea*, or taramea, *Aciphylla colensoi*, are particularly striking with their tall spires of yellow flowers, but, as their common name indicates, they are savage to forage and their edible roots and shoots are not regarded as delicacies. Also, some of their close relatives are at risk, so it isn't worth mistakenly uprooting those.

Portuguese angelica

SCIENTIFIC NAME
Angelica pachycarpa

OTHER NAME
Angelica

DESCRIPTION
Can grow up to 1m high, with thick, hollow, green stems which are sturdy and can be ribbed or furrowed. The plant has large, glossy, dark-green compound leaves with toothed or serrated edges that can grow up to 50cm long. Older plants produce large, rounded umbels, up to 15cm in width, with tiny, densely clustered flowers which are typically greenish-white. The plant has a slight aniseed smell.

DISTRIBUTION
Scattered across coastal locations throughout Aotearoa New Zealand.

HABITAT
Often found just above the high-tide mark in coastal areas, preferring frost-free environments with fertile and well-drained soils.

SEASONALITY
Available year-round, but produces fresh flowers mainly in spring and seeds in late summer and early autumn.

SIMILAR SPECIES
There are about five different angelica species available in Aotearoa. The larger garden variety, *Angelica archangelica*, is the one usually recommended for eating as it is less bitter than *Angelica pachycarpa*, but it is the latter you are most likely to find in the wild given that it has become an invasive weed.

Key edible uses

A very strongly flavoured aromatic plant, its bitter taste is most apparent in the roots and stems, but you can also eat the leaves and seeds. Cooking this angelica helps moderate the intensity of its flavour; just treat it as you would celery. You can eat the fresh stalks and leaves raw in fruit salads, or to season aromatic drinks, teas and cocktails, imparting a slight liquorice flavour. The leaves and stems provide a vivid-green colour when used as a fresh garnish or candied to decorate cakes and desserts. The seeds can be dried and used as a flavour boost for all sorts of dishes.

Opposite A fresh growth of Portuguese angelica on the bush floor. **Inset** The flowering umbel of the Portuguese angelica.

Wild chervil

SCIENTIFIC NAME

Anthriscus cerefolium

OTHER NAMES

Chaerophylum sativum, French parsley, garden chervil

DESCRIPTION

This green-stemmed garden escapee grows close to the ground, typically to a height of 30–70cm. It has very fine fern-like leaves that are bright green and delicate in appearance, and it produces tiny clusters of white flowers. It has a slightly resinous, aniseed smell.

DISTRIBUTION

Found in patches throughout Aotearoa New Zealand in low- to mid-altitude environments.

HABITAT

Prefers shady environments with damp and fertile soils; often found under oak trees in parkland environments or along the edge of pine and macrocarpa shelter belts.

SEASONALITY

Available year-round, flowering throughout spring to autumn with fresh growth in winter and early spring.

FORAGING TIPS

Can be easily damaged, so collect in a container when harvesting. Take care to remove any other green plants growing among chervil beds.

SIMILAR SPECIES

Take extreme care to identify chervil, as it looks very similar to the highly toxic hemlock (*Conium maculatum*; see page 46). Wild chervil has an aniseed aroma and green stems, while hemlock has an unpleasant odour. Hemlock is far taller and sturdier and its green stems are flecked with purple streaks or spots.

Key edible uses

Chervil is high in vitamins and minerals. The fresh green foliage is mainly used, though the white flowers make a tasty garnish and the roots are also edible. The leaves have a mild taste that is often described as a combination of parsley and anise. Chervil is a key herb in French cuisine, where it is added to bouquet garni, classic soups, sauces and salads. It can also be used for herb butter, mayonnaise, omelettes or to enhance vegetables. It is best used fresh or added at the end of a cook, as the flavour fades when the leaves are dried or heated.

Opposite Wild chervil grows well in shady, damp soils. **Inset** Wild chervil produces tiny, white flowers.

Cow parsley

SCIENTIFIC NAME
Anthriscus sylvestris

OTHER NAME
Wild chervil

DESCRIPTION
Grows in clumps with numerous stems up to 1m high. It has soft fern-like leaves, which are pale green to red, and large ornate white flowers with five petals that develop into seedheads in a radial cluster. Its flower stems are slightly hollow and furrowed, like celery. Because it has some red colouration on its stems, extreme care needs to be taken to distinguish it from hemlock (see page 46) with its red-spotted stems. However, the smell is a useful identification tool as, unlike hemlock, it has a pleasant, aromatic, carrot-like smell.

DISTRIBUTION
Patchy distribution in a wide range of locations near urban centres throughout the country.

HABITAT
Shady environments with damp and fertile soils, often growing under oak trees in parkland environments and on river margins.

SEASONALITY
Available year-round, flowering throughout spring to autumn, when it produces seeds. Fresh growth appears in late winter and early spring.

FORAGING TIPS
The leaves can be easily damaged, so collect in a container when harvesting.

SIMILAR SPECIES
Take extreme care to identify this plant as the leaves and flowers are very similar to the sturdier and darker-leaved hemlock (see page 46). Cow parsley has tiny hairs on its deeply ridged stems, unlike hemlock stems which are much smoother, more rounded and completely hairless.

Cow parsley could also be mistaken for giant cow parsley or giant hogweed (*Heracleum mantegazzianum*; see page 47), which has massive leaves and purplish stems and spots on its leaf stalk. When young, they are wildly different in size, with giant hogweed being very big.

Key edible uses

The fresh young leaves have the best aromatic, carrot-like flavour, but they can also be dried. The leaves make tasty additions to salads and are a useful parsley alternative. Some people consume the stems and roots, though the latter contain mildly toxic compounds so these should avoided, especially in pregnancy or if unwell. Like other plants, such as parsley, celery, citrus and figs, the sap contains compounds that cause photosensitivity in the skin, so it is best to protect against contact with the sap in strong sunlight.

Native celery

SCIENTIFIC NAME
Apium prostratum

MAORI NAME
Tūtae kōau

OTHER NAMES
Creeping marshwort, prostrate marshwort, sea celery

DESCRIPTION
Grows in clusters up to 70cm in diameter with long stems that are often smaller, thicker and lighter green than the similar Italian parsley. It is also similar in looks and taste to its relative garden celery (*Apium graveolens*). The white flower is very intricate, growing in tiny umbels off a central stem. The plants may exhibit yellow discolouration, which is due to a mineral deficiency.

DISTRIBUTION
Found throughout coastal New Zealand.

HABITAT
Favours rocky and stony shorelines just above the high-tide mark, particularly in areas with well-drained, calcium-rich soils. Often found in sheltered locations with overhanging shrubs.

SEASONALITY
Available year-round, but in best condition in winter and spring.

Opposite Native celery's intricate, white flower is just visible.

FORAGING TIPS
It is best to harvest the fresh growth and young outer stem — particularly avoiding the older stalks, which are more fibrous to eat. Avoid removing whole plants, as this native species is vulnerable to over-exploitation.

CAUTIONARY NOTES
If you have sensitive skin, be cautious about contact with the sap in sunlight.

Key edible uses

Its intense, salty, celery-like flavour means you might prefer to use it sparingly as a garnish or with other greens in smoothies and fruit drinks; however, the leaves and fresh stems are also useful in herb butter, salads, soups and stews or used as a parsley or celery substitute. High in vitamin C, it makes a good palate-cleanser. The flowers are also edible and make a visually appealing garnish. The seeds have a nutty, celery-like flavour and can be used as a spice to flavour meats, seafoods and many other dishes. The sweet, earthy root can be cooked and eaten like a vegetable.

Wild carrot

SCIENTIFIC NAME
Daucus carota

OTHER NAME
Queen Anne's lace

DESCRIPTION
The flowers are claret-coloured or pale pink before they open, then bright white and curving when in full flower. Some plants will have a dark, often purple dot in the centre of the umbel, distinguishing the plant from hemlock (page 46), which has similar flowers but no central spot. Growing no more than 1m tall, it has lush, finely structured leaves growing at its base.

DISTRIBUTION
Throughout much of Aotearoa New Zealand, especially in drier lowland areas, but largely absent from the west coast of the South Island and most higher-altitude areas.

HABITAT
Found in edge environments with very well-drained soils in regions with lots of sunshine.

SEASONALITY
Fresh growth in spring, flowers through the summer and goes to seed in autumn. Small amounts can be found year-round.

SIMILAR SPECIES
Take extreme care not to confuse with the much taller, extremely poisonous, hemlock (see page 46). Hemlock's smooth, hairless stem with its distinctive purple markings will distinguish it from the green, blotchless, hairy stem of the wild carrot. Each stalk within a hemlock's umbel holds flowers in loose clusters, while in the wild carrot the umbels are tightly packed and there may be that central dark dot. Note also the smell: hemlock's is unpleasant and musty; wild carrot, particularly its root, will (not surprisingly) smell of carrot.

Key edible uses

The whole plant can be used. The white root (carrot) is best roasted, or dried and ground as a coffee substitute, or grated to make a seasoning. The seeds also make a useful seasoning and an interesting addition to baking. The fresh young leaves are good in a salad or infused as a tea. Even the whole flowerhead can be used for flavour infusions or to make a stunning plating decoration.

Numerous claims are made for wild carrot's medicinal and diuretic properties, although conclusive research on any benefits — or indeed drawbacks — is yet to be done.

Opposite Wild carrot flowers in varying stages of bloom. **Inset** Fresh leaf growth of wild carrot.

Wild fennel

SCIENTIFIC NAME
Foeniculum vulgare

OTHER NAMES
Anethum foeniculum, Foeniculum foeniculum, Foeniculum officionale, sweet fennel, wild anise

DESCRIPTION
Often found in thickets, its feathery leaves are similar to the smaller dill. Its central stem is woody, green and hollow, reaching up to 1.7m in height. The vivid-yellow flowers are tiny, forming as clusters on a flat-topped umbel producing fine yellow pollen. The slightly elliptical seeds are also small, forming on the umbel 2–3mm in length. All parts of the wild fennel plant have a strong aniseed scent when crushed.

DISTRIBUTION
Most common in drier lowland locations along the east coast of Aotearoa New Zealand from Dunedin to Hawke's Bay, but also a weed around Northland and Auckland.

HABITAT
Common in well-drained soils in open sunny spaces, such as road and track edges, gravel riverbeds and wasteland.

SEASONALITY
Available year-round, with fresh growth in winter and early spring providing the best flavour. Flowers over the summer and seeds in the autumn.

FORAGING TIPS
A sharp knife is useful for cutting the firm fennel stalks.

CAUTIONARY NOTES
If you have sensitive skin, be cautious about contact with the sap in sunlight.

Key edible uses

All parts of the fennel plant are edible and have a pronounced aniseed flavour. The fresh leaf fronds are good in salads, added to seafood and chicken dishes, in tea or green smoothies where they match well with apple juice, and they can be candied in sugar to make a refreshing liquorice-flavoured treat. The firmer stems can even be used to make skewers. The roots of young fennel are straight and narrow (in contrast to the cultivated Florence fennel, *Foeniculum vulgare*), but can still be cooked and used as a root vegetable, especially when gathered after a period of rain. The pollen can be used as a spice, and it and the fronds and flowers all make excellent garnishes. The seeds, either when green or ripened to a greyish-brown, have a good flavour and work in a wide range of curries, stews, soups and aromatic dishes or in baking. The seeds can be sprouted to add to salads and sandwiches. The ripe fennel seeds can also be ground to flavour salt or be mixed with breadcrumbs into a seasoned coating.

Mountain carrot

SCIENTIFIC NAME
Gingidia montana

MĀORI NAMES
Kohepiro, koheriki, naupiro (note that these names are also used for other species)

OTHER NAMES
Angelica montana, mountain aniseed, New Zealand aniseed, New Zealand angelica

DESCRIPTION
Growing from a central base, this Aotearoa New Zealand native is a low-growing plant, reaching up to 30cm high and often well concealed among groundcover. It has robust stems with bright-green, jagged-edged, semi-succulent, glossy leaves that are shaped like pointed ovals and grow in pairs up the stem. Their white umbels are in loose clusters and produce dense seed clusters.

DISTRIBUTION
Widespread in high-country and alpine areas of the South Island and also found in higher parts on the ranges near Wellington, but now very rare in the North Island (where it was once more common and widespread). Still grows in reasonable abundance in the South Island.

HABITAT
Due to grazing pressure from introduced mammals, this species is now mainly found on rock crags and along steep banks; sometimes on lake edges. It likes well-drained sites and can grow in both shaded and open environments.

SEASONALITY
Most growth is in spring and early summer, with flowering from late spring through to early autumn. The plant tends to die back over the winter months.

FORAGING TIPS
Harvest carefully in small amounts, removing fresh-growth leaves without killing the plant. While not classified as endangered, it does appear to be vulnerable to grazing pressure. Harvesting seeds and growing in a rock-garden environment is also an option.

SIMILAR SPECIES
The loosely clustered umbels could be confused for hemlock (see page 46), but the foliage is quite different: thick rounded ovals rather than hemlock's fine ferny leaves.

Key edible uses

The leaves have an aromatic anise flavour and can be used as a garnish and palate-cleanser. The taproot and flowers can also be eaten. The seeds also have a strong aromatic flavour and can be used to make a flavoured salt.

Water celery

SCIENTIFIC NAME
Helosciadium nodiflorum

OTHER NAMES
Apium nodiflorum, fool's watercress

DESCRIPTION
Sprawling 0.5–1m high, water celery has creeping, hairless stems that root at the nodes in river sediments, meaning you'll often find the plants partially submerged in water. Its leaves are finely serrated and grow up the stem in pairs. The white flowers are arranged in umbels which radiate out from the base. While similar to watercress, it has longer serrated leaves.

DISTRIBUTION
Found throughout lowland North Island with strongholds in Wellington, Napier, Waikato and Auckland. For the South Island it is mostly found in Marlborough and Nelson and has recently spread to the Canterbury region. This introduced species is likely to continue increasing its spread.

HABITAT
Thrives in slow-moving freshwater environments, often in ditches and small creeks but also terrestrially in areas with damp banks and seepages.

SEASONALITY
Available throughout the year but tends to flower more in the summer and early autumn.

FORAGING TIPS
Gather fresh growth from the top of the plant. Only forage from locations with good water quality.

Key edible uses

Eaten across the Mediterranean, the leaves and young stems have a distinct carrot-like flavour. They can be added fresh to salads, pesto and smoothies, or steamed or boiled in a similar way to watercress. The white flowers are also edible with a strong, aromatic, celery-like flavour.

Opposite Water celery thrives in freshwater environments with slow-moving water.
Insert Water celery in flower.

Wild parsnip

SCIENTIFIC NAME
Pastinaca sativa

DESCRIPTION
A moderate-sized green plant reaching up to 1m high with green leaves coming off a central stem. It has a white root that grows to 30cm. The leaves are compound and lobed, with larger leaves at the base and smaller, more finely structured leaves higher up the plant. The pale-yellow flower umbel looks similar to fennel's but with a more robust structure.

DISTRIBUTION
Throughout Aotearoa New Zealand, but more common at lower altitudes although it can occur well inland in regions such as Central Otago.

HABITAT
Likes damp and nutrient-rich soils. Often found along the edge of roadside ditches or damp edge zones in rural environments.

SEASONALITY
Roots are best harvested in spring, with leaves picked in winter and flowerheads and seeds in spring and early summer.

FORAGING TIPS
Look for back roads through areas with rough farmland. The root is best harvested after a period of rain when the ground is soft; and preferably after frosts, which tend to bring a sweeter flavour.

CAUTIONARY NOTES
Be cautious about contact with the sap in sunlight, and keep the sap away from your eyes. The foliage often causes burns to the skin.

Key edible uses

Mainly harvested for the root, which often has a more intense flavour than cultivated parsnips and is high in minerals, vitamins, antioxidants and fibre. You can cook them just as you would a garden parsnip: baking or adding to stews and soups; you can also use it in cakes and puddings. The leaves and shoots can be cooked as a green, and along with the flowers and seeds they have an aromatic flavour and can be used in salads. The seeds can also be dried and ground to make a flavoured salt.

Above opposite The pale-yellow flower of the wild parnsip looks similar to that of wild fennel. **Below** A fresh harvest of wild parsnip roots.

Alexanders

SCIENTIFIC NAME

Smyrnium olusatrum

DESCRIPTION

Growing to roughly 1m high, alexanders has strong grooved stems and broad serrated leaves. The tiny green-yellow umbels hold flowers that are both male and female, with five long stamens raised above the pale-yellow petals. The mature fruit or seedpods are black.

DISTRIBUTION

Mainly in the Wellington region.

HABITAT

Wastelands and edge-lands, often in coastal areas, preferring damper areas but adaptable to most soils.

SEASONALITY

Flowers in summer, producing seeds in late summer; the fresh leaf growth is in autumn and the roots can be dug up in winter.

Key edible uses

Tasting like a pungent celery, the leaves, stems, flowers and taproot are all edible. The young leaves, shoots and stems can be eaten raw in salads or used as a parsley substitute. They can also be cooked along with the roots in stews, soups, etc. The flowers make colourful additions to salads, while the spicy seeds can be used like pepper.

Hemlock ⚠

SCIENTIFIC NAME
Conium maculatum

OTHER NAMES
Poison hemlock, wild hemlock

DESCRIPTION
All parts of this deadly plant are highly poisonous — medical assistance should be sought immediately if consumed as it can be fatal. With its umbels of tiny white flowers, hemlock looks like many other umbellifer plants, but the separate clusters making up each umbel are spaced out, not tightly packed. The dark-green, fern-like leaves begin as a rosette, but as the plant grows the leaves alternate up what can become a tall stem, usually over 1m and sometimes up to 2.5m high. The stems are hollow, often with purple spots or streaks, and while smooth, rounded and not grooved, the veins can look like ridges. It has no hairs on its stems or leaves. It also has an unappealing musty aroma, particularly when crushed, which some people compare to mouse urine.

Inset above and below The tiny white flowers of the hemlock make it very similar looking to other umbellifer plants. The key difference is that hemlock flowers are spaced out, rather than tightly packed.

Opposite above This plants gets its name from its sheer size — its massive umbels can grow up to 100cm. **Opposite below** Don't be fooled by the parsnip palm's attractive white-purple umbels. This is a very toxic plant.

Parsnip palm ⚠

SCIENTIFIC NAME
Daucus decipiens

OTHER NAMES
Melanoselinum decipiens, black parsley, Madeira giant black parsley, non-stinging hogweed, tree angelica

DESCRIPTION
This plant is poisonous and dangerous to touch. Growing up to 2m high, the parsnip palm has sturdy, woody stalks that are bare at the base, showing the horizontal lines of earlier leaf stems. Higher up, the stalk produces several bulky offshoots of large compound leaves with toothed edges. Both stalk and leaves are different to the umbellifers you are likely to forage, but it is the attractive 20cm white to purple umbels that might cause confusion. They produce dark-coloured seedheads after which the plant will die.

Giant hogweed ⚠

SCIENTIFIC NAME
Heracleum mantegazzianum

OTHER NAMES
Cartwheel-flower, giant cow parsley, giant cow parsnip, hogsbane, wild parsnip, wild rhubarb

DESCRIPTION
Similar to many of the umbellifers worth foraging, the sap of this plant will cause the skin to become sensitive to sunlight, but because of its size the hazard is increased and contact is very likely to cause first degree burns that will require medical treatment. Even brushing up against this plant should be avoided. It can grow as high as 5m with very large, sharply pointed, lobed leaves and large umbels up to 100cm. The hairy bright-green stalk is ridged with red spots.

Asparagaceae

Cabbage tree/tī kōuka, *Cordyline australis*

49

This is an unexpected family grouping, but these members share the common family attribute of significant roots or rhizomes. Star-like flowers appear in some of the species. Another diverse member of this small family is the agave, which is also edible — from its sap to its flowers, leaves and stalks — but as you're unlikely to find it in the wild it isn't covered here.

Rengarenga

SCIENTIFIC NAME
Arthropodium cirratum

OTHER NAMES
Maikaika, New Zealand rock lily, renga lily

DESCRIPTION
Grows to 70cm high, often found in clusters. The elongated strap-like leaves are a vibrant green with a soft texture. Clustering at the end of long stems, its white star-shaped flowers have six petals and open wide to display their stamens, which are yellow, white and pale purple.

DISTRIBUTION
Found throughout Aotearoa New Zealand, primarily in coastal areas, and especially in the upper half of the North Island.

HABITAT
In the wild they are found on rocky coasts, but they have been extensively planted for decorative landscaping in a wide range of environments, including parkland.

SEASONALITY
Available throughout the year.

SIMILAR SPECIES
Take care not to confuse with agapanthus, which has similar strappy leaves.

Key edible uses

The base of the fresh shoots can be eaten as a green when steamed. Traditionally the rhizomes were cooked in a hāngī, and they can be roasted with the fleshy stems.

Wild asparagus

SCIENTIFIC NAME

Asparagus officinalis

OTHER NAME

Sparrow grass

DESCRIPTION

Appearing from the bare ground as spears, wild asparagus has thinner, smaller stems than farmed asparagus. The colour of the spears can vary from green to purplish-green. If not harvested, the spears form a central green stalk, growing to almost 1m in height with very fine, hair-like, ferny leaves, which turn from bright green in spring to yellow in autumn. The flowers are tiny greenish-white bells, producing red berries in the summer (which are poisonous).

DISTRIBUTION

Throughout Aotearoa New Zealand but more common on drier South Island east coast locations.

HABITAT

Likes areas with well-drained silty or sandy soils, generally around coastal — even sand dunes — and lowland areas close to areas of human settlement, but they grow at some inland sites such as Alexandra.

SEASONALITY

New growth comes through in mid-spring, with late September to early November being the peak time to gather fresh edible tips. The plant flowers over the summer and fruits in summer.

FORAGING TIPS

It pays to mark out the plant's location in the summer when fully grown and conspicuous, and then return the following spring to look for the fresh young tips, which are often concealed. Older plants produce thicker stems, which can be woody though peeling makes them more digestible.

CAUTIONARY NOTES/CONFUSION SPECIES

The red berries are poisonous and must not be consumed. Note also that fresh growth of broomrape (*Orobanche minor*) grows from the ground as spears, so can be mistaken for asparagus; in fact one of its common names is wolf asparagus. Its flowers are very different once they develop, but the fleshy flower stalks, which vary between yellow-green and brown-purple, appear without foliage. The spearhead is hairy and feels sticky. There are varying accounts on whether it is toxic or not, but it is best not to try.

Key edible uses

A good source of vitamins and minerals, the fresh tips are often stronger in flavour than farmed asparagus. The tips can be eaten raw or stir-fried, grilled or steamed to bright green and served as a side dish. Once lightly blanched, the tips can also be preserved in a light vinegar solution to last through the year. The rest of the plant is inedible.

Cabbage tree

SCIENTIFIC NAME
Cordyline australis

MĀORI NAME
Tī kōuka

DESCRIPTION
A palm-like tree, growing up to 20m high, with a straight, brown-grey, corky-barked trunk. The stiff, angular, sword-shaped leaves are dull green and grow off upright branches in spiky mops. The tiny bright-white, star-shaped flowers are fragrant and produced on multiple branches coming off a main stalk sticking out from the heads of leaves.

DISTRIBUTION
Found throughout Aotearoa New Zealand but absent from high-altitude environments. Most common in coastal and lowland environments.

HABITAT
Adapts to a wide range of habitats, often found around wetlands and on hillsides. Can tolerate different soil types and moisture levels but likes growing in open, sunny locations and on the edges of bush. Also planted extensively as an ornamental tree in inhabited areas.

SEASONALITY
Spring for fresh shoots and flowers, and winter for roots. Flavours will vary according to rainfall and can become bitter after prolonged dry periods so best to harvest after rains.

FORAGING TIPS
A ladder is needed to access the top of trees. Trim carefully when you harvest so you do not risk killing the tree.

Key edible uses

Many different parts of the cabbage tree can be used. Young shoots at the base of the leaf clusters can be eaten both raw and cooked and have a spicy cabbage-like flavour. The fresh rhizomes can be dried out and pulped to release a sugary substance; they were a favourite for hāngī steaming but can also be put on an open fire. The flower stems have a mild nutty flavour. The flower buds are also edible, but are bitter when raw so best to be candied with brown sugar or dehydrated, which results in a mild aniseed flavour. The young shoots and inner core of the stem can be eaten raw or steamed, boiled or roasted, and have a slightly sweet and nutty flavour — discard the outer leaves and use the heart. The leaves can also be used to wrap food for cooking or as a substitute for banana leaves.

Hosta

SCIENTIFIC NAME
Hosta spp.

DESCRIPTION
A broad-leaved flowering plant. The leaves have distinct indented lines that curve from the rib of the leaf up towards the pointed tip. While there are varieties with differently coloured foliage in gardens, the ones you're likely to encounter in the wild are a bright light green. The pendulous flowers grow on long straw-like stems and are white or purple, opening into stars.

DISTRIBUTION
Mainly found in urban centres where the species is a garden escapee.

HABITAT
Found primarily in abandoned sections or near the edge of cultivated garden environments.

SEASONALITY
Fresh growth primarily in spring with the plant flowering over summer.

Above opposite Fresh hosta shoots. **Below** Purple hosta flowers on their long stems.

Key edible uses

The new shoots have an asparagus flavour and are good raw in a salad or cooked in stir-fries with the young leaves and crisp stems, which can be slightly bitter as they age; their flavour is best when lightly steamed. Chopping will overcome their potential to become fibrous, but young leaves make good wraps. The flowers are also edible, though use as a garnish as they are rather bland.

Asteraceae

Pineappleweed

The interesting feature of Asteraceae is highlighted by the various names it has gone by: the aster family, meaning star; the Compositae family, meaning composite; and the daisy or sunflower family, referring to its flowers. For each flower is actually many; a composite usually made up of a circle of petals radiating around a constellation of tiny flowers in the centre. If you think of the sunflower, the head produces loads of seeds — and that is because the central area inside the outer petals has multiple flowers.

A further point about sunflower seeds is that they are edible, and this is another common feature of this family. Many of its members are useful sources of food, such as lettuce or artichokes, or have medicinal uses, such as chamomile or marigold. Many members of this family can be bitter, usually because of their milky sap, but choosing fresh young plants, or crushing and rinsing older gatherings (as was done by early Māori), or boiling and discarding the water will make them more palatable.

It is easy to confuse a number of these species, as many have a similar daisy or tufted flower, as in the dandelion or thistle, while others have similar leaves. As most of the lookalikes are edible, confusion shouldn't be an issue — but, as usual, there are some members you need to know how to avoid, in particular the *Senecio* species, which all contain toxic alkaloids. Several species that are widespread in Aotearoa New Zealand, such as the low-growing groundsel (*Senecio vulgaris*), with its dots of yellow flowers hidden by green bracts, and ragwort (*Jacobaea vulgaris*), with yellow star flowers around a central disc, might initially look like edible species, so pay particular attention to the leaves, which will be key identifiers. Also, some people have allergies to the Asteraceae family, so in such cases consumption of plants in this family should be approached with due care.

Yarrow

SCIENTIFIC NAME
Achillea millefolium

OTHER NAMES
Devil's nettle, old man's pepper, woundwort

DESCRIPTION
Grows as a clump-forming herb with multiple erect, slender and branched stems arising from a basal rosette of leaves and reaching heights of 30–90cm. The leaves are fern-like with multiple small feathery leaflets. At first you might think the white or pink flowerheads are umbels, suggesting that the plant belongs to the Apiaceae family, but look more closely and you will see that each ring of petals surrounds composite flowers arranged in flat-topped clusters, which form numerous small brown seeds. The leaves and flowers have a strong aromatic scent. Once seeded, the plant often persists as a brown clump.

DISTRIBUTION
Throughout Aotearoa New Zealand but more common in dry regions. Absent from wet West Coast areas and high-altitude locations. One of our most common herbaceous plants in open dry environments.

HABITAT
Prolific in edge environments and rough pastures with sunny and open aspects. Tends to be most common in areas with well-drained soils, but will tolerate a wide range of conditions.

SEASONALITY
Available year-round but in optimal condition in spring.

FORAGING TIPS
Avoid collecting from busy roadsides, and wash in fresh water.

CAUTIONARY NOTES
While containing medicinal properties and while studies have shown that yarrow is generally safe to consume, extended use may cause skin allergies, and it is advised that yarrow, as an essential oil, should not be taken in pregnancy or if breastfeeding. It also contains the irritant thujone, so consume in moderation.

Key edible uses

All parts above ground are edible, though they can be bitter and the plant oils strongly flavoured, so use in small amounts. The leaves and flowers can be dried and used to make a refreshing tea or ground into a spice. When fresh, the young leaves can go into salads, soups, stews and other dishes as a leafy herb or garnish. The leaves can also be used in bitters and in beer-making.

Opposite These yarrow flowers are easily mistaken for umbels. Look closely however, and you'll see each ring of petals surrounds a cluster of flowers.

Lesser burdock

SCIENTIFIC NAME
Arctium minus

OTHER NAMES
Common burdock, button-bur, cuckoo-button, louse-bur, wild rhubarb

DESCRIPTION
In the first year of growth, the plant produces a rosette of wavy-edged, heart-shaped leaves close to the ground. In the second year, the leaves become more oval with a soft, cotton-like texture on the underside, and the plant sends up tall flowering stems up to 1.5m high. The flowers are pink to purple and form in globe-shaped clusters. They bloom in late summer and are known for their prickly heads, similar to thistle heads, which dry into spherical brown burrs with hooked spikes.

DISTRIBUTION
Found throughout the eastern and central South Island and the lower half of the North Island.

HABITAT
Generally found in edge zones with moist but well-drained soils and exposed to sunlight. Tends to establish in areas with some soil disturbance, such as roadsides, waste areas, abandoned fields and along the edges of forests.

SEASONALITY
Fresh growth starts in October and peaks through early summer with flowers in mid to late summer. A perennial species, so can be found year-round, but the roots are best harvested in early summer.

FORAGING TIPS
Use a hand trowel to extract the roots, and thoroughly wash off all soil fragments.

SIMILAR SPECIES
A few plants of *Arctium lappa*, or greater burdock, have started naturalising in Aotearoa New Zealand. It looks very similar to the lesser form, but is a lot bigger (up to 3m high). It can be consumed in much the same way (the nutritious root was used in the traditional drink 'dandelion and burdock' as well as in Asian dishes).

Key edible uses

The crisp, carrot-like roots have a mild, sweet flavour, contain mucilage so can be used as a thickener, and are a valuable ingredient in a wide range of Asian dishes, such as stir-fries, soups, stews and other recipes, and when roasted can make a coffee substitute. The young leaves and stalks have a tender texture and can be used as a salad green. Both the leaves and the roots become tougher and less edible in older burdock plants.

Common lawn daisy

SCIENTIFIC NAME
Bellis perennis

OTHER NAMES
Daisy, lawn daisy

DESCRIPTION
The leaves are typically spoon-shaped, with smooth or slightly toothed margins. They form basal rosettes close to the ground. The plant produces small, white or pinkish-white radiating petals around a yellow centre of composite flowers, typically 1–1.5cm in diameter.

DISTRIBUTION
Widespread throughout Aotearoa New Zealand.

HABITAT
Open sunny environments such as lawns, meadows, pastures and disturbed areas. They thrive in well-drained soils and are often considered lawn weeds.

SEASONALITY
Flowers throughout the year, with most flowering in springtime.

FORAGING TIPS
Pick from lawns with good soil quality and a spray-free environment.

SIMILAR SPECIES
The flower, though not the leaves and ground-hugging form, are very similar to the large range of plants of the *Erigeron* species, also known as fleabanes. These are also edible, see page 66.

Key edible uses

Flowers have a mildly sweet flavour and make a good plating garnish. The young leaves are mildly bitter and can be eaten raw in salads, sandwiches or added to soups or teas. Overall, this is an overlooked edible plant.

Calendula

SCIENTIFIC NAME
Calendula officinalis

OTHER NAME
Pot marigold

DESCRIPTION
An annual or short-lived perennial herb, growing 30–60cm high, with green leaves radiating out from a central base. The flowers are yellow or orange, with the typical daisy-like appearance, though the numerous ray florets surrounding the central disc may appear in several layers, depending on the variety.

DISTRIBUTION
Widespread throughout Aotearoa New Zealand as a common garden plant that occasionally grows in the wild.

HABITAT
Favours sunny, open areas with well-drained soils, generally in edge environments where the plant has escaped from gardens.

SEASONALITY
As the Latin name suggests in its reference to the calendar, the flowers are available year-round.

SIMILAR SPECIES
You could confuse the flowers with small zinnias or chrysanthemums, but these are also edible.

Key edible uses

The petals can be eaten fresh or dried; they have a pleasant flavour with a hint of pepper. They can add a bright taste to baking, herbal teas or soups, a splash of colour to salads, or act as an economical substitute for saffron in rice dishes or butter. The fresh young leaves are bitter but rich in vitamins and minerals and can be sprinkled in moderation into a salad.

Horseweed

SCIENTIFIC NAME
Erigeron sumatrensis

OTHER NAME
Fleabane

DESCRIPTION
A herbaceous perennial species growing from 0.5 to 1.5m high. It has a slender stem covered in fine hairs. The leaves are elongated, lance-shaped, and arranged alternately along the stem. The flowers are small and daisy-like, with white to pale-lavender petals surrounding a yellow centre. It flowers throughout summer and early autumn. The fresh-growth plant has a superficial appearance to tarragon.

DISTRIBUTION
Throughout Aotearoa New Zealand in drier regions, with a stronghold in Canterbury.

HABITAT
Open areas with dry, well-drained soils; often found along edges of farmland, parks and gardens.

SEASONALITY
Available throughout the year with fresh growth in early winter and early spring, flowering spring to summer with seeds from late summer through autumn.

Inset Fresh horseweed growth. **Opposite** Mature horseweed, turning to seed.

Key edible uses

The young leaves have a flavour similar to tarragon and are best dried and stored for use as a flavouring or as a pot herb. The young seedlings can be eaten fresh if cooked. When ground, the seedheads make a useful seasoning, with a flavour similar to onions.

Catsear

SCIENTIFIC NAME
Hypochaeris radicata

OTHER NAMES
False dandelion, flatweed

DESCRIPTION
A low-growing plant that forms a rosette of hairy leaves close to the ground. The leaves are lance-shaped and have toothed or lobed edges. The plant sends up one or more slender flower stalks from the centre of the rosette, each bearing a single yellow flowerhead. The flower resembles that of a dandelion (see page 84), with yellow ray flowers surrounding a central disc of tiny yellow disc flowers.

DISTRIBUTION
Widespread throughout the country.

HABITAT
Open environments with well-drained soils. Often grows in areas of low soil fertility.

SEASONALITY
Available year-round, with peak growth periods in spring and autumn.

SIMILAR SPECIES
To differentiate from dandelions, note the leaves: the toothed edges are much deeper in dandelions and not hairy. The stems on dandelions are wider, hollow and contain a milky sap. As dandelions are also edible, though, confusion is not an issue.

Key edible uses

Catsear has a deep taproot, which can be roasted to make a coffee substitute; fresh-growth roots can be lightly steamed. The young leaves are good in salad mixes or cooked, being mild in taste. The flowers are edible but tend to be very bitter, so use them sparingly.

Wild lettuce

SCIENTIFIC NAME
Lactuca virosa

OTHER NAMES
Acrid lettuce, bitter lettuce, opium lettuce

DESCRIPTION
Growing up to 1m high, this plant radiates out from a central point and looks similar to a tall dandelion. Leaves are reminiscent of its relative the garden lettuce (*Lactuca sativa*), but are greyish-green and have bristly margins and hairy undersides. The stems have a faint red colouration and carry milky sap. The dandelion-like yellow flowers are borne in clusters, similar to groundsel (*Senecio vulgaris*).

DISTRIBUTION
Found throughout drier parts of Aotearoa New Zealand. Absent from most of the South Island's west coast and alpine environments.

HABITAT
Open, dry and sunny environments. Often in areas of sand or gravel, as it likes well-drained soils. Occurs in a wide range of edge environments but largely absent from damp wetland areas.

SEASONALITY
Available throughout the year, but most growth is in spring and early summer.

FORAGING TIPS
Fresh growth is best harvested in spring.

CAUTIONARY NOTES/CONFUSION SPECIES
It is important to use only in small amounts, as the seedheads contains sedative compounds. There is insufficient research, but it may cause problems in pregnancy or if unwell or taking medication.

It is similar to *Lactuca serriola*, or prickly lettuce, and also sow thistle (*Sonchus oleraceus*; see page 82), both of which can also be eaten — but of more concern is the similarity of the closed flowerheads to those of the toxic groundsel *Senecio vulgaris*. Groundsel is shorter and its leaves quite different, being deeply lobed, so check carefully.

Key edible uses

Fresh-growth leaves are best to use in moderation added to a salad, or for wilting and serving as a side green. The flavour rapidly becomes bitter as the plant matures.

Nipplewort

SCIENTIFIC NAME
Lapsana communis

OTHER NAMES
Common nipplewort, dock cress

DESCRIPTION
An annual or biennial herbaceous species growing to 50cm high. It has spade- to arrow-shaped leaves, which can have paired leaflets at the base of the leaf stem, and notched edges. Long, delicate stems are multi-branched and hold the flower buds, which are green and caper-like in form, opening to pale-yellow petals around a centre with long stamens. It looks like a generally smaller, spindlier and simpler dandelion flower, with far fewer petals. Structurally similar to sow thistle (see page 82), but a smaller plant with finer growth.

DISTRIBUTION
Widespread throughout Aotearoa New Zealand except in heavily forested or alpine zones.

HABITAT
Typically likes areas of disturbed soils and is most common in edge zones. Will tolerate a wide range of soil types but prefers drier and well-drained soils. Avoids areas of acidic soils.

SEASONALITY
Available throughout the year.

FORAGING TIPS
Avoid collecting from busy roadsides.

Key edible uses

The radish-flavoured fresh-growth leaves and shoots are best collected before the plant flowers and then eaten fresh in a salad or wilted and used as a side green. When still green the flower buds can be pickled like capers. The yellow flower can be added to salads.

Inset Nipplewort looks very similar to dandelion, but with a lesser structure, and fewer petals.

Ox-eye daisy

SCIENTIFIC NAME
Leucanthemum vulgare

OTHER NAME
Chrysanthemum leucanthemum

DESCRIPTION
Grows to a height of 30–70cm with an unbranched stem that bears a solitary flower 2–7cm wide. The flower is composed of white ray petals surrounding a yellow central disc of tiny florets, like a large daisy. Leaves are simple, toothed and lance-shaped and form a rosette at the bottom of the stem.

DISTRIBUTION
A weed that is widespread throughout Aotearoa New Zealand, except in alpine and heavily forested areas.

HABITAT
Found in open environments with good sunlight and well-drained soils, often in edge zones or grassy or disturbed areas. Can grow in dense beds.

SEASONALITY
Available throughout the year, but leaves are best foraged in late winter and early spring with fresh-growth flower buds through spring and summer. Dies down over winter.

CAUTIONARY NOTES/CONFUSION SPECIES
Some people may experience mild skin irritation from contact with the leaves and flowers; a feature that is more pronounced in related species which should not be foraged, though they are less common in the wild in Aotearoa. For instance, the shasta daisy *Leucanthemum* x *superbum*, has a mildly toxic sap that can cause skin and stomach irritation; its daisy flowers are much bigger than its cousin, often a palm's width. Its parent *Leucanthemum maximum* is probably also best avoided; again it is larger than the ox-eye.

Key edible uses

Highly nutritious, this daisy has a slightly bitter flavour which intensifies as the plant ages, so stick with new leaves and flowers, which can be eaten raw in salads and sandwiches or cooked in omelettes or stews. Finely chopped, the young shoots and root can also be eaten when cooked. The green flower buds can be pickled and eaten as capers, and the flowers dried to make tea. Consume in moderation.

German chamomile

SCIENTIFIC NAME

Matricaria chamomilla

OTHER NAMES

Matricaria recutita, camomile, false chamomile

DESCRIPTION

A branching plant growing up to 60cm high. It has bright- to medium-green feathery leaves that are finely divided and arranged alternately along the stems. The most distinctive feature of chamomile is its small (1.5–2.5cm) daisy-like flowers, which have a yellow central cone of florets that is markedly raised above the usual circle of white petals, which seem to drop back a little. The flower has a sweet, apple-like fragrance — a key identification characteristic.

DISTRIBUTION

Sparsely dotted around urban centres across Aotearoa New Zealand.

HABITAT

Primarily a garden escapee into modified environments, such as park tracks and edge zones around urban areas. Likes well-drained soils and sunny environments.

SEASONALITY

Flowers mainly from October to January.

CAUTIONARY NOTES/CONFUSION SPECIES

Those with ragweed allergies should avoid chamomile. Stinking chamomile *(Anthemis cotula)* is very similar, but has a distinctive, unpleasant odour; its acrid juice is mildly toxic, causing skin and stomach irritation.

SIMILAR SPECIES

Anthemis nobilis, also known as *Chamaemelum nobile*, is the English or Roman or 'true' chamomile, and being slightly smaller and more sprawling in habit is often used to make lawns. Although it is a different species, it is very similar to German chamomile and can be used in the same way. Being smaller, though, it is less generous to the forager than its German counterpart.

Key edible uses

Chamomile tea, with its pleasant, herbal, apple-like flavour and mildly sedative properties, is a well-known use of the dried flowers, which can also flavour a variety of dishes and beverages, such as salads, soups, stews and desserts. The flavour can be quite delicate, so shines where it can be infused into syrups, sauces, vinegars, cordials or alcohol. The dried flowers will also impart their distinct flavour when added to dough or batter, such as shortbread or cakes, or to infuse cream for ice cream, or when added to honey to sweeten teas or cocktails, or when drizzled over yoghurt or oatmeal, or used in a glaze for roasted vegetables. The aromatic leaves can be consumed in all the same ways, though they have a milder flavour.

Pineappleweed

SCIENTIFIC NAME
Matricaria discoidea

OTHER NAMES
Disc mayweed, rayless mayweed, wild chamomile

DESCRIPTION
A low-growing green plant with a fine, ferny leaf. Its central disc is raised and greeny-yellow, but unlike its German cousin (*Matricaria chamomilla*, see page 72) its outer petals are so minimal you'll think there are none.

DISTRIBUTION
Found throughout Aotearoa New Zealand in dry and mainly frost-free areas.

HABITAT
Often found in sunny open areas on the edges of gravel roads and tracks.

SEASONALITY
Available from September to April with a peak from December to February. Tends to die away over winter.

CAUTIONARY NOTES
People with allergies to plants in this family should not consume this plant.

Inset Pineappleweed flowers have a lovely, sweet pineapple scent.

Key edible uses

The flowerhead when bright yellow has a strong pineapple flavour and can be used raw in salads or for infusing flavours into syrups and vinegars. The flowers can also be dried to make herbal tea. Fresh-growth foliage can also be added to salads.

Wall lettuce

SCIENTIFIC NAME
Mycelis muralis

OTHER NAMES
Lactuca muralis, Prenanthes muralis

DESCRIPTION
Typically grows in sprawling clusters from rosettes with wide and deeply lobed, sometimes red-tinged leaves (a bit like exaggeratedly indented dandelion foliage), the end of which is a large pointed arrow-shape. It has loose clusters of small (5mm), all-yellow, ray-like flowers, made up from what look like five scalloped petals though each is in fact a separate flower. They produce seeds like dandelions (see page 84).

DISTRIBUTION
One of our most widely spread wild greens, occurring throughout the country, including alpine environments, but is rare north of Auckland.

HABITAT
Often at the edge of tramping tracks in both open and forested environments as well as rocky conditions. Often grows on walls and rocky surfaces.

SEASONALITY
Available year-round but less available over the winter months.

Key edible uses

Its sharp, bitter flavour makes its young leaves a vibrant addition to salads or as a cooked vegetable green. Can also be used to make pesto.

Mouse-ear hawkweed

SCIENTIFIC NAME
Pilosella officinarum

OTHER NAMES
Hieracium pilosella, hawkweed

DESCRIPTION
A low-growing plant, up to 40cm high, with leaves radiating out from a central point and a lemon-yellow dandelion-like flower on a long slim stalk. There is sometimes a faint red stripe on the back of the petals. Unlike a dandelion (see page 84), its leaves, which generally grow flat on the ground, are oval with a pointed tip, grey-green and covered in long hairs. There are 10 similar species in both the *Pilosella* and *Hieracium* genuses, collectively known as hawkweeds; this one is distinguished by each plant producing only a single flower and its hairy leaves.

DISTRIBUTION
An invasive weed, widespread throughout drier parts of Aotearoa New Zealand.

HABITAT
Areas of open spaces with dry or well-drained soils, infiltrating grassland and tussock zones.

SEASONALITY
Available year-round with a flush in spring and a secondary flush in autumn.

Key edible uses

Traditionally this nutrient-dense plant has been used medicinally but it is not highly rated for its culinary value, though it is useful when mixed with other species, the fresh-growth leaves and flowers being added fresh to salads or dried to make herbal tea.

Variegated thistle

SCIENTIFIC NAME
Silybum marianum

OTHER NAMES
Marian thistle, milk thistle

DESCRIPTION
A tall herbaceous plant with a robust and prickly appearance, growing to 1m high. The spiny leaves are large, glossy and deeply lobed, radiating from a central base, and are green with a distinctive white marbling and prominent milky-white veins. The typical thistle flowers are purple and bloom in clusters at the ends of thick, sturdy stems covered with fine white hairs. Each flowerhead is composed of numerous individual florets surrounded by spiky bracts. The seeds are encased in a spiky, round or oval-shaped seedhead, which is light to dark brown. The roots are sturdy and fibrous.

DISTRIBUTION
Generally found in lower to mid altitudes in the drier parts of eastern Aotearoa New Zealand, although it can be found some distance inland in the lower North Island. Tends to have a more patchy distribution than the more common Scotch thistle.

HABITAT
Prefers dry, open areas where there has been some ground disturbance; often found on stony river margins and agricultural edge environments.

SEASONALITY
Available year-round but best foraged during wetter periods in spring and autumn when there is more likely to be fresh growth. Tends to die back after flowering in late summer.

FORAGING TIPS
At their best when gathered in spring when soft. Use gloves when handling and scissors to cut off the sharp edges.

CAUTIONARY NOTES
Do not gather where nitrates may be high, such as fertilised areas, as the plant can absorb high levels. As with all plants in this section, do not consume if you are allergic to the Asteraceae family.

SIMILAR SPECIES
You can also forage the new growth of the familiar Scotch thistle. It can be eaten in a similar way, plus the bland tap roots can be cooked like a parsnip or eaten raw.

Key edible uses

The young stalks, leaves, roots and flowers are mildly flavoured and can be eaten raw or cooked. The sharp spikes of the leaves must be removed and the central stem peeled before being consumed in salads, as a side green, or like asparagus or rhubarb. The unripe flower buds can be cooked or pickled. The roasted seeds can be ground to make a herbal drink or used as a coffee substitute.

Sonchus species

With small dandelion flowers and dandelion-like leaves, the sow thistles foraged in this country look very much like taller, rangier, multiple-flowered versions of their well-known cousin. Putting aside the dandelion similarity, though, there is a fair amount of further confusion over sow thistles and the name pūhā. This is partly because, as with so many of the Asteraceae family, there are a number of lookalikes; though conversely the same species can vary in appearance, changing with age or according to where they grow. And there is also confusion because the name pūhā is used for several different *Sonchus* species, not just one. To add to the confusion, plants with Māori names are usually indigenous; however, pūhā is now widely used for introduced sow thistles, the name presumably transferring from native to similar plants when it was discovered they were just as edible. And not just edible but tasty and nutritious, as sow thistles are high in vitamins and minerals, in particular vitamins C and A, calcium, potassium and iron.

There are over a hundred sow thistle or *Sonchus* species across the world. Although it is commonly said that there is only one native species out of four in Aotearoa New Zealand, there are actually more. Native sow thistles, though, are out of bounds. Two plants bearing the pūhā name are at risk: *Sonchus grandifolius*, also known as Chatham Island sow thistle, embergeria and pūhā pārākau rahi; and *Sonchus kirkii*, also known as New Zealand sow thistle, prickly sow thistle, shore pūhā or raurōroa. *Sonchus grandifolius* is quite distinct, so won't be mistaken for other sow thistle species, and *Sonchus kirkii* may have similar flowers but its much larger, grey-green, hairless leaves, while toothed, are not as indented or spiny as the pūhā you want to forage. Another native, *Sonchus novae-zelandiae*, is commonly called dryland sow thistle and mostly grows in the central South Island. It is threatened so should also be left alone, but again despite the familiar flower its leaves bear only a glancing similarity to other sow thistles, so it should be easy to exclude from your foraging radar.

The sow thistles or pūhā to target as a forager are all introduced species. Two are included here, but there is also *Sonchus arvensis*, the perennial sow thistle, which also has the name raurōroa. With larger flowers than the other introduced sow thistles, which are annuals, it has appeared around Christchurch and on the Chathams, in particular, but is less likely to be encountered than the two below, though it can be foraged and eaten in the same way. Just be aware that sow thistle roots can absorb nitrates, so do not forage in areas of intensive fertilisation.

Above opposite Fresh growth sow thistle, with closed buds. **Below** Thistles have unique, toothed leaves.

Prickly and common sow thistle

SCIENTIFIC NAME

Sonchus asper, sonchus oleraceus

MĀORI NAME

Pūhā, rauriki

OTHER NAMES

Hare's colwort, milky tassel, rough milk thistle, soft thistle

DESCRIPTION

Growing over 1m high, prickly sow thistle is generally twice as high as the common sow thistle, but both have an erect stem often branching near the top. The pointed leaves can be deeply lobed and irregularly toothed or lance-shaped, containing white sap. They may look prickly in the common species but are soft, while the prickly sow thistle has sharp spines under the leaves and along the margins. The yellow dandelion-looking flowers are borne in clusters at the ends of the stems producing dandelion-like seedhead 'clocks'.

DISTRIBUTION

Found throughout Aotearoa New Zealand but absent from higher-altitude areas, above 1000m.

HABITAT

Found in a wide range of habitats, including fields, waste areas, roadsides and disturbed sites. It thrives in sunny locations and can tolerate a variety of soil types.

SEASONALITY

All year round.

FORAGING TIPS

Gather foliage when young and before flowering, as it will be less bitter.

Key edible uses

The leaves can be eaten raw in salads or smoothies, or boiled or steamed as a green side dish or added to pesto, stews and soups. They have a slightly bitter taste, similar to other wild greens, which moderates when cooked or if you discard the mid-ribs. When peeled and rinsed of sap, the stems can be added to the boil-up or eaten like celery or asparagus. The green flower buds will pickle well like a caper, and when open can be salted and added to salads. The roots can be roasted and ground as a coffee substitute, but are smaller than dandelion roots (see page 84), which are preferable.

Opposite Pūhā is a term for many different thistles in Aotearoa New Zealand. **Inset** Pūhā in flower.

Common dandelion

SCIENTIFIC NAME
Taraxacum officinale

MĀORI NAME
Tawao

DESCRIPTION
Grows from a rosette of deeply lobed leaves of 4–8cm in length. Unlike sow thistles (page 82), their stems do not branch and the leaves do not grow off the ground, but they do have similar forms and toothed edges to pūhā.The bright-yellow tufty flowers open from green buds and are composite, consisting of numerous small florets which go on to form the spherical 'clock' seedheads.

DISTRIBUTION
Widespread but more common in lowland locations.

HABITAT
Areas of well-drained short grass with moderate levels of moisture. Likes open sunny environments such as parks and lawns.

SEASONALITY
Available year-round but growth peaks in spring and autumn, especially after periods of rainfall. Tends to die away over summer and winter.

FORAGING TIPS
Best to harvest new growth before flowers appear and after a period of rain and in moist locations, as the flavours can be very bitter following dry periods or on older plants. A trowel is useful for extracting the roots.

CAUTIONARY NOTES/CONFUSION SPECIES
Avoid collecting from lawn areas where sprays are used.

Note that there is a native species of dandelion, *Taraxacum gilliesii*, also known as *Taraxacum magellanicum*, Gillies dandelion and tohetaka. It is a little smaller but very similar to *Taraxacum officinale*; the good news is that it is not at risk and is also edible.

Avoid eating the stem as it is high in latex.

Key edible uses

Most of this weed can be consumed, and it is rich in antioxidants, several different vitamins and minerals, and may have medicinal qualities as well, though more research needs to be done. Fresh green leaves can be added to a salad or pesto or sautéed in olive oil and garlic, or boiled and served as a side green. Boiling will help to remove the bitterness, so is suggested for older leaves. The flower buds when firm and green and nestled among the leaves, can be pickled like capers or cooked. Once the flower has opened, the petals can be plucked from the flowerheads and added to a salad or used to flavour vinegars, teas or to make wine. Its large nutritious roots can be boiled like a slightly bitter root vegetable, but are traditionally slow-roasted and ground to make a nutty, caffeine-free coffee.

Salsify

SCIENTIFIC NAME

Tragopogon porrifolius

OTHER NAMES

Jerusalem star, oyster plant, purple salsify, vegetable oyster

DESCRIPTION

A biennial species growing to 1m high with narrow, lance-shaped leaves. It is easy to spot when it produces its striking star-shaped flowers that radiate out with green pointed bracts behind spiky petals, which are typically purple, and a centre of tiny florets. Salsify has a long, slender and tapered root. The root can grow to 30cm long and has a thin brown outer skin that can be easily peeled to reveal a creamy-white inside. Salsify produces seedheads similar to dandelions (see page 84), with a fluffy, spherical 'clock'.

DISTRIBUTION

Mainly drier east coast sites throughout Aotearoa New Zealand.

HABITAT

Likes open environments with sunny aspects and well-drained soils, such as roadside margins and sometimes dry edges of riverbeds. Often found along coastal margins and will tolerate soils with low salinity.

SEASONALITY

The main flowering period is September to January, with the roots best harvested from mid-autumn to mid-winter, a frost reportedly improving its flavour.

FORAGING TIPS

Mark out the plants' location when flowering and return over the cooler months to harvest the root.

Key edible uses

The primary culinary attraction of salsify is its long tapering nutritious root, which has a mild, slightly sweet flavour and a tender texture when cooked like other root vegetables, including boiling, roasting, frying or adding to soups and stews. If young, the root can be grated and eaten raw in salads. The new green shoots, leaves, flowers and sprouted seeds are also edible, although seldom used.

Opposite A bed of salsify at Purau Bay — Lyttleton Harbour, Banks Peninsula, going to seed. **Inset** A fresh, spiky salsify flower.

Berberidaceae

Oregon grape

Berberis species are in the Berberidaceae family, and Aotearoa New Zealand has varieties that have arrived from Europe, South America and Asia. Their leaves usually spiral up the stems, which have two forms: the main long stems and the short stems which grow from them. Their edible berries contain iron, various other trace elements and a number of vitamins, in particular vitamin C, making them excellent targets for foraging. They are readily spread by birds and possums, so getting in first will reduce their tendency to become invasive. In fact, *Berberis darwinii* is even banned for sale in this country and must not be propagated. The only drawback to harvesting these plants is their other common feature: thorns.

Alongside other beneficial compounds the berries contain berberine, an antioxidant used in medicine to treat a variety of ailments; as with any ingredient containing distinct chemicals, use in moderation and avoid giving to young children or consuming while pregnant or while sick, especially if taking any medication. Note that drying the berries will concentrate the levels of berberine.

In addition to the species described here, you might see others in gardens, also producing edible berries. For example, Oregon grape, *Berberis aquifolium*; Japanese mahonia, *Berberis japonica*; *Berberis soulieana*; Japanese barberry, *Berberis thunbergii*; and Wilson's barberry, *Berberis wilsoniae.*

Darwin's barberry

SCIENTIFIC NAME
Berberis darwinii

DESCRIPTION
An evergreen shrub, growing to 3m high with glossy, dark-green, oval leaves, which are stiff and edged with spaced-out spines, reminiscent of holly, and spiny branches. The glowing orange-yellow flowers appear in prolific dangling racemes or clusters, each small bud beginning bell-shaped before the outer petals open to form what almost look like tiny orange daffodil heads. They produce edible berries that are dark blue-black when ripe.

DISTRIBUTION
Found mostly from Rotorua southwards, particularly in drier eastern regions but absent from higher-altitude areas, above 500m. Most prolific in the Wellington and Dunedin regions.

HABITAT
Adapts to most soil types, often colonising scrub in open environments with good exposure to sunlight, particularly rolling hill country, forest margins or in coastal areas.

SEASONALITY
Flowers in spring with fruit in summer and autumn.

FORAGING TIPS
Having a ladder can be useful for accessing berries at the top. Take care to avoid thorns.

SIMILAR SPECIES
Despite its name, the great barberry *(Berberis glaucocarpa)* is actually smaller than its cousin but is very similar, its leaves slightly more elliptical, lighter green and with more distinct veins. Its fruit is also edible and blue-black with a powdery white bloom, as is often seen on grapes or plums veins.

Key edible uses

Full of nutrients, the fruit has a strong, tart, slightly sweet flavour: the riper the berries, the less sour they are. They can be used for a wide range of dishes requiring a hit of acid or where you would normally add orange peel. Barberries can be eaten raw, made into jellies and jams (they are a useful source of pectin), added to chutneys and pickles, dried and added to muesli, or candied. They can also be juiced for making smoothies, syrups, flavoured spirits or cordials (straining out the seeds). The juice is a rich purple colour and makes striking vinegars or infusions.

Opposite Darwin's barberry often forms dense thickets. **Inset** Flower.

Mountain grape

SCIENTIFIC NAME

Berberis oiwakensis

OTHER NAMES

Mahonia lomariifolia, *Mahonia oiwakensis*, Chinese hollygrape

DESCRIPTION

This Asian barberry is more of a tree than its shrubby cousins, growing up to 7m high with woody, branch-like stems, which are often bare apart from the top where the leaves spring out in a crown. These leaves are large — up to 45cm long — but divided up their stem into smaller, well-spaced, paired, lance-shaped leaflets. In comparison with the other barberry species, the margins of these leaflets show only a hint of spaced, spiny notches. The bright-yellow, belled flowers are produced up the length of multiple spires, usually rising up from the top leaves. Later in the season, these spires will droop down from the weight of the multiple oval purple-black berries. Like *Berberisglaucocarpa* (see page 90) the berries have a white bloom and the plant also possesses thorns.

DISTRIBUTION

This barberry originated in areas of high elevation, but in Aotearoa New Zealand it seems to have adapted to both mountainous and coastal conditions and is dotted patchily across the country, usually in urban areas.

HABITAT

Currently this plant is found in margin areas, having not strayed far from gardens and parks.

SEASONALITY

Flowers through late winter and spring and produces its berries in early summer.

FORAGING TIPS

Gloves and a ladder will make harvesting easier.

Key edible uses

The nutritious, tart berries can be eaten raw or cooked. Being juicy, they make excellent drinks and syrups, but they can also be dried or candied and used as a cranberry substitute.

Clockwise from left The woody, branch-like stems; mountain grape in flower; mountain grape fruit clusters.

European barberry

SCIENTIFIC NAME
Berberis vulgaris

DESCRIPTION
Growing up to a bushy 2–4m high, this species also displays savage thorns up its stems. The spines on its more rounded, oval leaves are much closer together than on the other *Berberis* species, giving a toothed appearance. Fresh growth is green but as the fruit appears the foliage turns yellow-red, before dropping over winter. The yellow bell-shaped flowers open to look a bit like buttercups, appearing on usually much shorter racemes than its barberry cousins. It is the berries, though, that are the biggest contrast: clusters of bright crimson-red berries, a bit like rosehips.

DISTRIBUTION
More common in Canterbury, though also found in urban areas, particularly around Dunedin, Nelson and Wellington, with a few appearances further north.

HABITAT
Having been used for hedges, it is evident in urban margins.

SEASONALITY
Flowers in the summer, producing berries in autumn.

FORAGING TIPS
Use thick gloves to avoid the spines.

Key edible uses

Eaten both raw and cooked, the nutritious, tart berries, being high in pectin, are good for making jellies and jams, but also for adding to syrups, chutneys and pickles. When dried they are excellent alternatives to cranberries or currants, and they can also be candied. Traditionally, they are included in a range of Persian and Uzbek dishes, such as pilaf, stews and soups, or for stuffing chicken or trout; in India they brighten vegetables or go into desserts; the Germans make barberries into a sweet red-wine sauce; the French make a marmalade with the berries; and in Russia they are used for drinks and candies.

Brassicaceae

Twin cress seedlings often grown among rocks

From leaves to flowers, to seeds, to roots, the members of this family are entirely edible, and none will poison you. Most are strongly flavoured, which isn't surprising given that mustard is a member, as is horseradish (*Armoracia rusticana*), which is rare to find in the wild in Aotearoa New Zealand, but is out there. As well as heat, this family also has a tendency to bitterness, especially when the plant is older, which frustratingly is when you can be sure of identification.

Brassicas can be identified by their small — usually white, yellow or pink — flowers, which always have four well-spaced petals; most also have six stamens. Previously this family was called Cruciferae, referring to their cruciform or cross-shaped flowers. There are other plants that have only four petals but are not part of this family, though, so don't rely solely on this for identification.

The leaf shapes vary, but they often grow in a rosette form from a central base. The flowering stems produce seeds in capsules, which often make good eating when they are still crisp and green. These may range from the tiny heart shape of shepherd's purse to the oval discs of honesty to the elongated capsules of wavy bittercress or pea-like pods of wild radish.

Many of our cultivated vegetables are from this family, and you might find them making a bid for freedom in urban edge environments or close to farmland. For instance, rapeseed (*Brassica napus*) is usually seen creating startlingly bright-yellow fields on farms, where it is grown for its oil, but it is starting to be found in the wild and will give you fresh shoots and leaves for salads, along with mustard-flavoured seeds. Cabbage, broccoli and other *Brassica oleracea* species, or rocket (*Eruca vesicaria*), should all be familiar from the garden plot if you encounter them when foraging; other plants will be obvious once you dig down, such as turnip (*Brassica rapa* subsp. *rapa*). These cultivated varieties will have different characteristics from the wilder forms, the breeding aimed at increasing size or improving taste. Yellow-flowered *Brassica fruticulosa*, for instance, known as twiggy turnip or Mediterranean cabbage, is actually a wild form of cabbage but its leaves are more like those on broccoli. It might be found wild, particularly in Wellington and Christchurch, but, while edible, it has mostly been used for traditional medicine in Sicily and certainly won't be producing a cabbage head.

There are a number of native members of the Brassicaceae family, such as coastal cress, *Lepidium flexicaule*, Kawarau cress, *Lepidium sisymbrioides*, or scurvy grass, *Lepidium oligodontum*, but as these are endangered you should only be aware of them so that you can ensure you leave them alone.

Land cress

SCIENTIFIC NAME
Barbarea verna

OTHER NAMES
Early yellow rocket, upland cress, winter cress

DESCRIPTION
A large-leaved cress with glossy deep-green leaves made up of leaflets, often growing close to the ground with a yellow flower.

DISTRIBUTION
Throughout New Zealand but more common in South Island's east coast.

HABITAT
Often found on gravel riverbeds and disturbed gravel road edges. Likes cooler and well-drained areas.

SEASONALITY
Available year-round but with a main peak in spring and a lesser peak in autumn.

Key edible uses

Treat like watercress: cook or eat raw in salads or sandwiches, particularly the young sprouts. Like all brassicas, the yellow flowers are edible too. Land cress has a similar hot flavour to watercress. The seed can be used to create oil (if harvested in bulk).

Above insert Land cress in flower.
Below Fresh leaf growth.

Sea rocket

SCIENTIFIC NAME
Cakile spp.

OTHER NAMES
American searocket, Harper's searocket, European searocket

DESCRIPTION
A low-growing green-coloured coastal shrub that can be found both in bushy patches and as single plants. It has a dense succulent texture, producing oval green seedpods from purple flowers (*Cakile maritima*) and white flowers (*Cakile edentula* though this can also come in pink, blue and purple shades).

DISTRIBUTION
Both species are widespread throughout coastal eastern regions and tend to be more widespread in the North Island.

HABITAT
Sand dunes, often on the frontal edge, just above the high-tide mark.

SEASONALITY
Available in all stages throughout the year.

FORAGING TIPS
Wash sand off. Can be grown in a pot with sandy soil in frost-free environments.

Inset Sea rocket's unique seedpod.

Key edible uses

All parts of this plant can be eaten. It has a tangy, bitter flavour, so the leaves can be blended and made into a thick paste as a foraged substitute for wasabi. The new leaves are milder so can be sprinkled into salads, while the older ones are best cooked. The leaves and pods can be pickled to moderate the hot bitterness. The purple flowers make a good garnish, especially with meat dishes, and add some spicy flavour. The root can be ground and used as a flour additive for making bread. Extremely hot/bitter, so use in moderation.

Hairy bittercress

SCIENTIFIC NAME

Cardamine hirsuta

OTHER NAMES

Flick weed, lamb's cress, shot cress, spring cress

DESCRIPTION

A small, low-growing cress that grows out radially with small, four-petalled, white flowers and hairy leaves.

DISTRIBUTION

Throughout Aotearoa New Zealand.

HABITAT

Grows in areas of disturbed soil and often among wood chips. Prefers damp and shaded locations along disturbed edge environments.

SEASONALITY

Available year-round but peaks in abundance over late winter and spring and is less common over the summer months.

FORAGING TIPS

A delicate plant, so place in a container after foraging to prevent it from getting crushed.

SIMILAR SPECIES

Very similar to wavy bittercress, *Cardamine flexuosa* (see page 101), and there is a further *Cardamine* species you might encounter, though it is less common: the purple-flowered *Cardamine pratensis*, commonly known as cuckooflower or lady's smock. There are also a couple of native species, which are not endangered and are also edible — *Cardamine dolichostyla*, or panapana, and *Cardamine forsteri* — but being so small they are not really worth foraging.

Key edible uses

The flowers and leaves are useful as a garnish or in salads. While it can be cooked, it is best eaten raw as the cress-like flavour is delicate, though as the name suggests it is slightly bitter.

Wavy bittercress

SCIENTIFIC NAME
Cardamine flexuosa

OTHER NAME
Wood bittercress

DESCRIPTION
A low-growing plant with bunches of small, deeply lobed, bristly leaves. It produces clusters of white flowers with delicate, wavy stems and thin seedpods.

DISTRIBUTION
Found throughout Aotearoa New Zealand, generally in lower-altitude areas.

HABITAT
Moist and sheltered areas such as woodlands, damp farmland, riverbanks and the edges of lawns.

SEASONALITY
Growth peaks in spring with a smaller spurt in autumn. Tends to die away over the winter months.

CONFUSION SPECIES
Similar to hairy bittercress (*Cardamine hirsuta*; see page 100).

Key edible uses

The leaves and young stems can be eaten raw. Leaves have a mildly bitter, cress-like taste and are often used as a peppery addition to salads or as a garnish. Leaves can be used in soups and stir-fries. The peppery flavour makes it useful in small amounts in pesto.

Shepherd's purse

SCIENTIFIC NAME
Capsella bursa-pastoris

DESCRIPTION
A small delicately structured plant with serrated leaves radiating out from a central base, heads of small white flowers and heart-shaped seedheads on long stalks (the purse).

DISTRIBUTION
Widespread throughout drier parts of Aotearoa New Zealand.

HABITAT
Likes areas with disturbed well-drained soils, and is found in both urban and agricultural edge zones, in open environments with good amounts of sunlight. Often found in wastelands.

SEASONALITY
Available year-round but leaves are best foraged in the spring.

FORAGING TIPS
Strip the seeds off the central stem.

Opposite A bed of shepherd's purse in flower. **Inset** Shepherd's purse has a distinct heart-shaped seedhead.

Key edible uses

High in beneficial vitamins and minerals, the fresh leaves and young flowering shoots can be eaten raw or cooked and have a mild mustard-like flavour. The seeds are ideal for sprinkling through a salad. The white flower has a mild cress flavour and is a good garnish. You can also make a tea from the dried plant.

Sweet alyssum

SCIENTIFIC NAME
Lobularia maritima

OTHER NAME
Sweet alison

DESCRIPTION
Low-growing flowering plant with clustered small white or purple flowers and long, spindly green leaves. Grows in tight clusters.

DISTRIBUTION
Widespread throughout coastal regions but prefers drier areas often close to urbanised locations.

HABITAT
Fast-growing, white, scented flowers often planted in gardens as groundcover, or in borders and pots, but also naturalising between paving stones and in disturbed areas. Prefers well-drained soils.

SEASONALITY
Year-round.

FORAGING TIPS
Flowers can be collected in clusters. Wash thoroughly and gently in a salad spinner.

Inset Sweet alyssum grows in clumps often along garden edges. **Above opposite** Sweet alyssum flowers are often white but some become purple. **Below opposite** Wild alyssum in bloom.

Key edible uses

Leaves, stems and flowers can be used in salads to add pungency. Becoming increasingly popular as a readily available and tasty garnish flower.

Twin cress

SCIENTIFIC NAME
Lepidium didymum

OTHER NAMES
Lesser swine-cress, swine wartcress

DESCRIPTION
Grows at ground level as a creeping plant with sprawling, clumping foliage with small leaves and clusters of tiny green seedheads.

DISTRIBUTION
Throughout coastal and lowland Aotearoa New Zealand but can occur at inland sites.

HABITAT
Well-drained soils, often growing in gravel or even out of cracks in concrete. Likes areas where the soil has been disturbed.

SEASONALITY
Available throughout the year, but in best condition in late winter and spring.

FORAGING TIPS
Cut off fresh growth and wash thoroughly.

CAUTIONARY NOTES
Because it is so low-growing, take extra care to remove any grit from the foliage.

SIMILAR SPECIES
There are other *Lepidium* cresses that you might come across and can then forage, but their spread is currently limited compared with twin cress. Examples are: common peppercress, *Lepidium africanum*, Argentine cress, *Lepidium bonariense*, swine cress, *Lepidium coronopus*, bushy pepper cress, *Lepidium desvauxii*, heart-podded hoary cress, *Lepidium draba*, and garden cress, *Lepidium sativum*.

Key edible uses

Initially has a distinct savoury flavour that evolves into an intense heat, so makes a hot paste to replace wasabi. Small amounts of fresh growth in a salad will spice it up. Can also be added when stewing vegetables or indeed fruit to provide some spicy heat; it goes particularly well with stewed apricots.

Clockwise from left Twin cress with seeds; seedling; a bed of twin cress in flower, with tiny green seed heads.

Honesty

SCIENTIFIC NAME
Lunaria annua

OTHER NAMES
Moneywort, moonpennies, penny flower, silver dollar

DESCRIPTION
A tall, leggy plant, growing up to 90cm high, with the usual four petals of the brassica family, which are bright pink-purple or white, depending on the variety. Large arrow-shaped leaves are hairy and have toothed edges. The seedheads begin as green circles, which become papery and browny-white when they mature, the outer layer falling off to reveal the seeds and leave the distinctive transparent coins.

DISTRIBUTION
An escapee from floral gardens, it has spread throughout Aotearoa New Zealand, particularly around urban areas.

HABITAT
An adaptable plant, though it prefers moist, shady or dappled conditions, so is often found in shrubby wasteland or the edges of parks and bush.

SEASONALITY
Flowers in summer, with seeds appearing in autumn.

FORAGING TIPS
Collect the leaves in spring, as they become bitter when the flowers are produced.

Key edible uses

The leaves have a mustard flavour and can be eaten raw in salads or cooked as a side green. The seeds are horseradish-flavoured. When unripe, the green seedpods can be eaten fresh as instant snacks or used to give a kick to salads. When ripe, the seeds can be ground to make a version of mustard, becoming stronger when mixed with cold water and left to sit for a quarter of an hour. For a milder, bitter version, mix with hot water or vinegar. The root can be eaten raw or roasted.

Stock

SCIENTIFIC NAME
Matthiola spp.

OTHER NAME
Gilly flower

DESCRIPTION
A genus of various species and cultivars, which have escaped the flower garden. A showy plant with spires of strongly scented four-petalled flowers in pale pink, pink-purple or white, depending on the variety. The leaves are lance-shaped ovals and usually grey-green and floppy.

DISTRIBUTION
Found in coastal areas close to the main urban centres.

HABITAT
Preferring chalky and sandy soils, it is often found on coastal cliff-tops.

SEASONALITY
Flowers in summer, with seeds appearing in autumn.

Above inset Purple stock flowers. **Below** A bed of stock with white flowers.

Key edible uses

The sweet and slightly peppery flowers will add colour and fragrance as a garnish or in salads or desserts. The seedpods can be eaten but are not particularly palatable.

Narrow-fruited watercress

SCIENTIFIC NAME
Nasturtium microphyllum

OTHER NAME
Onerow yellowcress

DESCRIPTION
Trailing along waterways, the bright-green new leaves turn purply-brown after flowering. The oval lobes growing up the leaf stem have a peppery scent. Its tiny white flowers cluster at the top of the hollow stalks, looking quite similar to those of shepherd's purse (see page 102). It can be easily confused with watercress (see page 114), but, as suggested by its common names, its pods hold the seeds in a single row so are longer and much thinner.

DISTRIBUTION
Found across the country but mainly close to areas of settlement.

HABITAT
An aquatic plant, it can be found on the edges of flowing fresh water and in damp gullies.

SEASONALITY
Flowers from spring to summer, which is the best time to pick the leaves.

CAUTIONARY NOTES
Watercress easily absorbs pathogens and toxins, so ensure that the waterway is not contaminated by pollution or industrial activity, and that there are no livestock nearby — parasites such as liver fluke can be passed on; if in doubt, cook your foraging as this will kill any parasites. Consume soon after processing to prevent the watercress from being exposed to bacteria.

Key edible uses

Particularly rich in vitamins, antioxidants and minerals such as iron, the hot-flavoured leaves can be eaten raw or cooked. As with watercress, the fresh raw leaves can be used in pesto, salads, sandwiches or herb butters. When cooked, they are good in boil-ups, fry-ups or steam-ups. The flowers can be eaten as a garnish and the seeds blended with salt to make an aromatic seasoning. Alternatively, the ground seeds can make a mustard substitute, adjusted to your taste: when mixed with cold water a pungent flavour will develop after about 15 minutes; when mixed with salt, vinegar or hot water, you will get a bitter but milder taste.

Opposite A bed of narrow-fruited watercress showing typically more slender leaves than the common watercress.

Watercress

SCIENTIFIC NAME
Nasturtium officinale

MĀORI NAME
Kōwhitiwhiti

DESCRIPTION
Grows in dense mats of bright green. Leaf shape is variable depending on the age of the plant: rounded when young but progressively elongated. Has clusters of small, four-petalled, bright-white flowers with six yellow stamens. Forms a narrow green seedpod.

DISTRIBUTION
Throughout the country in freshwater environments, from lowland to alpine regions.

HABITAT
Grows anywhere where there is fresh water. The best growths often occur in cold flowing water along small stable watercourses, but also commonly found in roadside drains and freshwater backwaters.

SEASONALITY
Available year-round, but growth is more prolific in spring. Dies back over winter. Flowers can be gathered from September to April and are in peak season over the summer months.

FORAGING TIPS
Best gathered from close to the source of a spring creek when eating fresh, as watercress easily absorbs pathogens and toxins. Use a sharp knife to cut off the top foliage and leave the roots to allow regrowth. Use waders.

CAUTIONARY NOTES
Avoid collecting from catchments with livestock nearby, as parasites such as liver fluke can be passed on — if in doubt, cook your foraging as this will kill any parasites. Also avoid watercress growing in areas with industrial activity, as it can absorb toxic metals.

Key edible uses

Particularly rich in vitamins, antioxidants and minerals, this is a versatile mustard-flavoured green. The fresh raw leaves make a very good pesto base, can be added to a salad, sandwiches, herb butter or used as a garnish. They can be cooked in boil-ups and soups, sautéed in olive oil, steamed in hāngī or used as a substitute for pūhā. When cooked, the distinct peppery flavour is moderated, and the leaves go a dark-green colour. The flowers and seeds can be blended with salt to make an aromatic seasoning. Alternatively, the ground seeds can make a kind of mustard: when mixed with cold water a pungent sulphurous flavour will develop after about 15 minutes; when mixed with salt, vinegar or hot water, a bitter but milder taste will result.

Wild radish

SCIENTIFIC NAME
Raphanus raphanistrum

OTHER NAME
Charlock

DESCRIPTION
A long-stalked plant with lobed, toothed leaves growing in a rosette form. Their four-petalled flowers are either yellow or white with distinct darker veins, and form narrow pea-like seedpods. Unlike the garden radish, the root does not become a globe but instead is branched.

DISTRIBUTION
Generally, in coastal areas with a stronghold in the Wellington region.

HABITAT
Well-drained and fertile soils often along disturbed coastal margins but can grow inland and in gully systems.

SEASONALITY
Available year-round, but most growth in spring with seedpods in summer.

FORAGING TIPS
Best to collect after some rain. Seedpods can become tough in texture after prolonged dry periods or if old. Flowers are delicate and need to be put in a container when collected.

CAUTIONARY NOTES
Often grows in areas of dumped soil, so be aware of eating the roots.

SIMILAR SPECIES
You may also find the pink-flowered, round-rooted garden variety *Raphanus sativus* gone rogue, or, more rarely, the yellow-flowered Mediterranean radish, *Raphanus raphanistrum landra*. All are worth foraging if you come across them.

Key edible uses

The whole plant can be used. Roots can be harvested during wet conditions. The fresh leaves can be added to stir-fries or shredded and put in a salad to add spicy heat. Older leaves are more bitter but useful in ferments, such as kimchi and sauerkraut. The flowers are handy garnishes (although with a short shelf-life), and when still buds can be treated like broccoli and briefly steamed. When fresh and young, the raw seedpods have a pleasant flavour and crisp texture. The seeds are much more pungent when dried and either pickled or ground and used as a wasabi paste. Collected seeds can be processed for their oil (the plant is related to oilseed rape).

Clockwise from left Distinct four-petalled flower in white; yellow flower; fresh growth leaves.

Bog yellow cress

SCIENTIFIC NAME
Rorippa palustris

MĀORI NAME
Poniu

OTHER NAMES
Marsh cress, marsh yellowcress

DESCRIPTION
Grows up to 80 cm high, but often close to the ground with a sprawling form. Leaves are oblong with toothed margins, varying from light to dark green. It produces small yellow flowers on tall stems.

DISTRIBUTION
Throughout Aotearoa New Zealand at both coastal and inland locations.

HABITAT
Likes sunny open aspects on the edges of wetland areas, especially along the water edges and in ephemeral wetland. Will often grow in areas of fine silt or sediment.

SEASONALITY
Main season is spring to late autumn, often dying away over the winter months, although may be found at some northern locations growing all year round.

CAUTIONARY NOTES
Bog yellow cress has been known to accumulate high levels of nitrates, particularly in areas of agricultural runoff, so it is recommended to consume it in moderation and only from clean water sources.

Key edible uses

This cress has edible leaves that can be consumed raw or cooked. The leaves have an aromatic, spicy taste and are often used as a salad green. Larger leaves can be wilted as a side green. It has a similar flavour to watercress but is sweeter. The yellow flowers are also edible, so are useful as a garnish.

Opposite Bog yellow cress grows well in sunny wetland areas. **Inset** Bog yellow cress in flower.

Hedge mustard

SCIENTIFIC NAME

Sisymbrium officinale (formerly *Erysimum officinale*)

DESCRIPTION

Medium-sized plant with long, lobed leaves with pointed ends, growing up to 70cm high with distinct pale-yellow flowers and a wiry seedhead.

DISTRIBUTION

Found throughout Aotearoa New Zealand in low- to mid-altitude areas. Most common in edge environments around urban centres.

HABITAT

Prefers areas with well-drained soils. Often found growing in among rank grass on the edges of tracks, roads and streams.

SEASONALITY

The greens are best foraged in winter and early spring, with flowers appearing through late spring to mid-summer. Tends to die away over the autumn months.

FORAGING TIPS

Wash thoroughly as often found on the edges of tracks and roads.

SIMILAR SPECIES

Found in Christchurch and Dunedin areas in particular, *Sisymbrium orientale* is known as oriental or Indian hedge mustard or eastern rocket. It is similar to *Sisymbrium officinale* and can be eaten in the same way.

Key edible uses

The new growth of leaves has a pleasant, slightly bitter, mustard flavour of a medium heat. The leaves can be used in salads, ferments or cooked together with the stems. The yellow flower makes a good garnish with a wide range of savoury and meat dishes. The seed can be consumed raw or cooked, and ground to add flavouring to soups and stews or as a wasabi replacement. It was used as a medicinal herb by colonial settlers to treat digestive ailments and throat complaints. Recent research has identified various compounds that suggest antioxidant, antimicrobial and anticancer properties, though further work is needed to confirm these findings.

Nasturtium

SCIENTIFIC NAME
Tropaeolum majus

OTHER NAMES
Cardamindum majus, Tropaeolum elatum, Tropaeolum hortense, Trophaeum majus, Indian cress, monk's cress

DESCRIPTION
As its looks suggest, this nasturtium is not in the Brassicaceae family but rather part of Tropaeolaceae, but both families are members of a wider order known as Brassicales. The presence of mustard oils in Brassicales indicates the family connection, and the hot spicy flavour that comes from it is the reason it has been included in this section. This is a slightly aromatic, ground-creeping, trailing plant with large circular leaves growing up to 15cm across, and bright yellow, orange or even red flowers. It has a rounded, ridged green seed.

DISTRIBUTION
Widespread throughout coastal Aotearoa New Zealand as a garden escapee, so often found around or near areas of human settlement.

HABITAT
Prefers frost-free areas with a high amount of sun exposure and well-drained and sandy soils.

SEASONALITY
Available year-round in most locations.

CAUTIONARY NOTES
Although highly nutritious, children may dislike the peppery heat of this plant.

Key edible uses

High in vitamin C and other beneficial properties, all parts of the plant have a distinctive peppery kick. The young leaves are good eaten raw in salads, pesto and salsas or used as mini wraps. Larger leaves can be blended to make a flavoured oil. The flowers are also edible and, as with the leaves, make striking garnishes; both can be added to stir-fries. The immature seedpods provide an alternative to capers when pickled, while the ripe seeds can be ground into a pepper substitute.

Opposite Nasturtium can be seen throughout the country, in gardens and in the wild. The peppery flowers come in a diversity of bright colours, and the leaves have many culinary uses.

Caryophyllaceae

Sea campion

This family is also known as pinks, referring to the colour of the familiar garden flower dianthus, or carnation. Their edible, clove-scented petals can be used as garnishes, an ingredient for syrups or candying and a colourful addition to both savoury and fruit salads. Having edible flowers is a feature of several other members of this family, which have naturalised in this country.

Sea campion

SCIENTIFIC NAME
Silene uniflora

DESCRIPTION
A low-growing perennial herb that grows in thick mats to 30cm high. It has grey-green leaves and produces solitary flowers on slender stalks. The flowers have 5 divided petals (so they look like 10 petals) and are usually white or pale pink, protruding from a bell-shaped calyx which is distinctly veined.

DISTRIBUTION
Mainly in coastal Canterbury and Wellington with a scattering of other coastal locations around the country.

HABITAT
Well-drained and open environments along coastal margins, often on sand dunes and in among rock walls.

SEASONALITY
Fresh foliage in late winter and spring and flowers in summer. Often dies away over the early winter months.

CAUTIONARY NOTES
Silene spp. do contain quite high amounts of saponins so it is best to cook them, which reduces the amounts. It is a small-leaved plant however, so it is unlikely to be collected in any great quantity. A few raw flowers decorating a salad are not going to cause gastric upset. Saponins are part of a plant's natural defence and are present among other everyday food items.

SIMILAR SPECIES
Bladder campion (*Silene vulgaris*) is a slightly taller species (up to 50cm), which is very similar though generally has more flowers per stem and more delicate petals. The fresh-growth leaves and shoots are eaten in the Mediterranean with plenty of olive oil in salads and risottos and along with the petals can be consumed in the same way as sea campion.

Key edible uses

The leaves and fresh young shoots are sweet and so can be eaten raw in salads, but they become increasingly bitter with age, which can be moderated if lightly steamed, boiled or fried. The leaves can be eaten as a side dish or added to stews or omelettes. The petals separated from the calyx are also edible.

Inset and opposite Sea campion's distinctive flowers have five petals, divided to look like they have 10.

Chickweed

SCIENTIFIC NAME

Stellaria media

DESCRIPTION

Small, delicate leaves that are oval-shaped, growing from 1 to 3cm long. The stems are thin with fine hairs growing down one side, and often form dense mats, spreading across the ground. When you break the stem, it will have a white, elastic lining. Grows to a height of 20–40cm. A delicate plant that crushes easily and has a spindly structure. The plant produces tiny white flowers, with 5 divided petals which look like 10, forming little stars as their scientific name suggests.

DISTRIBUTION

Throughout Aotearoa New Zealand, except alpine environments and areas that have heavy frosts on open landscapes. More common around areas of human habitation.

HABITAT

Found in a wide range of edge environments, especially gardens, fields and wastelands. Likes sheltered areas with some shade and medium-to-high soil fertility.

SEASONALITY

Throughout the year with fresh growths following periods of rainfall. Can die away quickly in cold or dry conditions.

FORAGING TIPS

Often a range of other greens will grow among patches of chickweed, so take care to remove any foreign plants.

CAUTIONARY NOTES/CONFUSION SPECIES

Check that your foraging area is spray-free. The plant contains saponins, which mostly pass through the body and are not a problem for healthy people, but are mildly toxic. The saponins are reduced through cooking. We consume them in our everyday diets in such foods as quinoa, oats and spinach, even eating a salad bag full of this plant would be fine.

There are various other plants called chickweed in Aotearoa, including several native ones; only *Stellaria parviflora* has similar leaves though its petals are barely visible.

Key edible uses

A good source of vitamins and minerals, the leaves, stems, flowers and seeds are all edible, imparting a mild, slightly earthy flavour. Usually eaten raw in salads, pesto and green smoothies, but can be cooked in stir-fries or stews if added at the last minute. Also makes an ornate garnish.

Clockwise from left Chickweed has a tiny, white flower, and the stems have fine hairs growing down one side; the leaves, stems and star-like flowers of this plant are all edible.

Ericaceae

Alpine heather in bloom in Tongariro National Park

This is the heather family, whose members are mostly evergreens. Ericaceae includes small edible heath plants, such as blueberries and cranberries; but also much larger shrubs and trees, such as rhododendrons, which are poisonous. Blueberries are not included here as they have naturalised only recently and not spread far as yet, but keep an eye out as they may soon start seeding further afield, judging by the way heather has taken off in the Central Plateau.

Speaking of heather (*Calluna vulgaris*), it may not seem particularly edible with its small, needle-like foliage, arranged in pairs along woody stems, but don't overlook it. The small, bell-shaped flowers, borne in dense, terminal clusters usually in pinkish-white, can be used to flavour beverages such as heather ale or as an aromatic seasoning in alcoholic botanicals. Flowering sprigs — fresh or dried — can be simmered to make herbal tea, and the foliage has also been used to smoke and flavour meats, such as game. And as it is fast adapting to our hill country, it won't hurt to forage it.

There are a lot of native Ericaceae species in Aotearoa New Zealand — some at risk, many thriving — though only a few are of interest to the forager. The most tasty are included here, but, given that these are heath plants, most can only be found in remote areas so their primary culinary use is to provide a fresh snack on a long tramp.

Strawberry tree

SCIENTIFIC NAME
Arbutus unedo

OTHER NAME
Cain apple

DESCRIPTION
Typically reaching heights of 5–8m, this evergreen tree has a rounded crown and reddish-brown bark. The dark-green serrated leaves are narrow ovals ending in points. The tree produces drooping clusters of small bell-shaped white flowers with pink stalks, which bear unique berry-like fruit. Dangling like pompoms from the branches, the fruit starts yellow and ripens into an orange-red with a bobbly surface that looks a bit like a strawberry but with a tougher texture.

DISTRIBUTION
Found throughout Aotearoa New Zealand but absent from higher-altitude areas above 500m. Tends to be more common around areas of habitation with limited distribution into the wild.

HABITAT
Found in open and edge environments, mostly in urban centres, in parkland and garden environments, and occasionally planted as an ornamental roadside tree. Likes areas with strong sunlight and low rainfall.

SEASONALITY
Flowers appear in autumn, the fruit ripening from March to June.

FORAGING TIPS
Having a ladder can be useful for accessing fruit higher up on the tree, though the fruit will fall when ripe.

Key edible uses

The bright fruit are high in antioxidants and omega-3 and -9 fats and can be eaten raw, providing a crunchy outer texture and a soft yellow flesh that has a mild, sweet flavour. The ripe berries can also be placed in sodas or alcohol for flavour infusions. They can also be cooked into jam.

Snowberries

SCIENTIFIC NAMES

Gaultheria antipoda, *Gaultheria depressa*, *Gaultheria macrostigma*

MĀORI NAMES

Koropuka, pāpapa, takapo, taupuku, tāwiniwini

OTHER NAMES

Bush snowberries, mountain snowberries, prostrate snowberries

DESCRIPTION

Gaultheria antipoda is a shrub that can grow to 2m high, while the other two species are low-growing shrubs, in particular the prostrate *Gaultheria macrostigma*. They all have small, thick, oval, green leaves, white bell-shaped flowers and large white, pink or red berries, which can be as big as 10mm, made up of about four joined lobes.

DISTRIBUTION

Gaultheria antipoda is spread throughout the country, though the other two do not tend to go much further north than the Central Plateau. Mostly found in regions of higher altitude (above 300m) among low shrub and widely spaced tussocks.

HABITAT

On hill- and mountain-sides, but occasionally in areas of open, flat, scrubby environments in higher-altitude areas. Likes lots of sunlight and often found on the edge of tracks.

SEASONALITY

The peak season for berries is from February to April, but they are juicier harvested early.

FORAGING TIPS

Look carefully at ground level for larger white snowberries, which are often well concealed.

CONFUSION SPECIES

Do not confuse with ornamental snowberry, *Symphoricarpos albus*, is often grown as a garden ornamental, as it will cause vomiting. Its leaves are noticeably larger than the fruit (in *Gaultheria*, the leaves and berries are similar in size with the fruit sometimes being larger). Also, the *Symphoricarpos albus* berries are not lobed but smooth.

Key edible uses

Berries are usually eaten raw and have a mild, sweet-floral flavour with the larger puffy white snowberries being the most tasty. You can include them in jams and pies.

Opposite Large, white snowberries typically grow close to the ground.

Prickly mingimingi

SCIENTIFIC NAME
Leptecophylla juniperina

OTHER NAMES
Cyathodes juniperina, inangapōriro, miki, ngohungohu, taumingi, tūmingi

DESCRIPTION
A prickly small to medium-sized shrub that can grow up to 4m high, looking a bit like tea tree, *Leptospermum scoparium* (see page 184). The leaves are narrow and sharply pointed, typically arranged in sets of three, giving the plant its 'prickly' appearance. The shrub produces small white-to-pink flowers, which are usually solitary and resemble a bell that opens at the bottom into a five-pointed star. The flower produces small white-to-pink, sometimes dark-pink, berries containing tiny seeds.

DISTRIBUTION
One of the most widespread mingimingi species in Aotearoa New Zealand, although absent from some drier and alpine regions.

HABITAT
Often grows in edge environments, particularly on forest margins or among scrub on hills, but also in subalpine areas.

SEASONALITY
Fruits throughout the year but with a peak from February to April. Foliage is available throughout the year.

Key edible uses

The berries can be eaten and have a mild flavour. The foliage can be dried and used for making herbal teas or included in a mix for smoking foods.

Pātōtara

SCIENTIFIC NAME

Styphelia nesophila

OTHER NAMES

Leucopogon fraseri, dwarf mingimingi

DESCRIPTION

Typically less than 10cm high, this low-growing alpine plant forms dense patches with stems covered by whorls of tiny, hard, angular, sharp-tipped leaves. Like prickly mingimingi (page 136), its white tubular flowers open at the bottom into a five-pointed star. Attached to the stem, sometimes in clusters, are small, rounded, orange berries, 3–5mm in size.

DISTRIBUTION

Found throughout Aotearoa New Zealand, often in subalpine areas.

HABITAT

Likes sunny areas with open environments and low vegetation. Often found in windswept locations and, while primarily subalpine, can also be found in coastal herb fields.

SEASONALITY

Fruits from January to April.

Key edible uses

The small orange berries are juicy with a mild, sweet flavour and can be eaten raw on the spot. Looking like salmon roe, as a garnish they will add a bright splash of colour to the plate.

Fabaceae

Gorse flowers

This family, also known as the Leguminosae or legume family, includes well-known edible plants such as peas, beans, lentils and chickpeas, so unsurprisingly it also has wild varieties of interest to the forager. The seeds are generally high in protein and essential amino acids; however, the wild forms are not necessarily always edible. Indeed, there are many plants in this family that are particularly toxic to humans, such as all types of broom, kōwhai, laburnum, lupins and wisteria, so view it with caution and take note that while one part of a plant may be safe to eat, another may not be. Spanish broom in particular is highly toxic, and can be easily confused with the more common European broom. Spanish broom in particular is highly toxic, and can be easily confused with the more common European broom.

You can identify a member of Fabaceae by the fruit, which comes in the form of a legume, or seamed pod, containing the seeds: think of the typical bean pod. Also its leaves are usually compound, meaning they have separate leaflets attached to the main rib; and often their flower has the distinctive asymmetrical, five-petalled blossom typified by the sweet pea.

Sydney golden wattle

SCIENTIFIC NAME
Acacia longifolia

OTHER NAMES
Acacia sophorae, coastal wattle, golden wattle, long-leaved wattle

DESCRIPTION
A shrubby tree growing up to 8m high. What look like lance-like leaves are actually adapted stalks, or phyllodes, which are flattened grey-green blades rather than foliage. The flowers are yellow and narrow, looking like 3–4cm-long bottlebrushes. The prolific seeds are held in what look like long flattened bean pods that can grow up to 12cm but sometimes coil.

DISTRIBUTION
Mostly Christchurch northwards, particularly in Auckland and from Wellington up the west coast to Whanganui.

HABITAT
Invading wasteland, bush, dunes and pastures, this species prefers sunny lowland positions.

SEASONALITY
Flowering in spring, the seeds are produced in summer.

CAUTIONARY NOTES/CONFUSION SPECIES
Although there are a number of wattle species that are edible, there are some that contain alkaloids and so are toxic. There is conflicting information about the edibility of the other wattles commonly found in this country (silver wattle, *Acacia dealbata*, black wattle, *Acacia mearnsii*, and blackwood, *Acacia melanoxylon*). Although some but not all of these may be toxic, their yellow flowers look like fluffy balls rather than bottlebrushes, making them easy to confuse. It is safest to stick with the long-flowered *Acacia longifolia*.

Key edible uses

The new leaves, flowers and seeds are edible cooked or raw. Remove the stalks from the flowers and use in baking, in egg dishes or as fritters. With a nutty, slight coffee flavour, the seeds are high in protein and traditionally were dried, roasted and ground into flour. They can have a slightly sulphurous flavour if eaten unroasted. These days wattle seeds are used in a variety of ways, in both sweet and savoury dishes. The sweet, sticky gum is also edible.

Opposite The bright yellow flowers are a dazzling highlight on the landscape in the spring.

Broad-leaved sweet pea

SCIENTIFIC NAME
Lathyrus latifolius

OTHER NAMES
Everlasting pea, perennial pea

DESCRIPTION
A climbing plant that uses tendrils to support itself as it grows, allowing it to scramble over fences, trellises, and other vegetation. It has laterally compressed stems, and oval leaflets produced in pairs with soft, curved tendrils. Flowers are a mix of white, pink and purple, which are borne in clusters. It has a small thin seedpod.

DISTRIBUTION
Has a patchy distribution throughout Aotearoa New Zealand, found mainly throughout drier and eastern parts and at some inland sites in Otago. However, it also grows in scattered sites throughout the North Island.

HABITAT
Grows as a creeping vine over a range of low-growing vegetation or on slopes or over rocks, often near areas of settlement. It likes open sunny areas with well-drained soils in a range of edge environments. Typically found alongside roads, train tracks and water-courses but will venture into coastal edge zones.

SEASONALITY
Tends to die back over winter with fresh growth from late winter and spring and flowers from September to March, with seedpods in mid to late autumn.

CAUTIONARY NOTES/CONFUSION SPECIES
There is a lot of debate about the toxicity of sweet peas, and there are related species, notably *Lathyrus sativus*, also known as Indian pea, grass pea or chickling pea, that are the focus of many toxicology studies, though *Lathyrus odoratus* — the garden sweet pea — is often cited too. Certain chemicals contained in all *Lathyrus* plants can lead to lathyrism, an irreversible neurological disease. However, even in *Lathyrus sativus*, the seeds have to be eaten in large quantities over a sustained time. An occasional small meal of *Lathyrus* peas is unlikely to be poisonous, but it is safer to stick to the flowers. Some people claim these are also poisonous, but research identifies the seeds and pods as the primary source of toxicity.

Key edible uses

The fresh-growth tendrils are collected in spring and have a crunchy texture and pleasant pea flavour, and can be used as a garnish or in salads. The flowers make a pleasantly flavoured garnish on a wide range of dishes and have become popular in restaurants. It is best to avoid the seeds and pods.

Opposite and inset Purple sweet pea flowers are typical but white flowers also occur. The broad-leaved sweet pea often grows in dense beds, as pictured here.

Black locust tree

SCIENTIFIC NAME
Robinia pseudoacacia

DESCRIPTION
A medium-sized deciduous tree growing to 10–15m tall in Aotearoa New Zealand. Its crown is broadly rounded; its leaves consist of 9–19 leaflets going up their stalk in pairs and ending with a final single one. Each leaflet is elliptical, 3–5cm long and has a smooth edge. The tree produces fragrant, white, pea-like flowers that hang in drooping clusters up to 20cm long. The brown elongated pods are slightly curved and contain several seeds.

DISTRIBUTION
Widespread but patchy throughout Aotearoa New Zealand.

HABITAT
Often occurs in parkland areas, with limited dispersal into the wild. Also often found in planted river margins close to areas of human habitation.

SEASONALITY
The tree produces ripe flowers from mid-September until early November.

CAUTIONARY NOTES
The leaves, bark and seeds contain toxalbumins, which are toxic to humans. They do break down with heat, which may be why some sources say you can eat the seeds and pods — but it is best to stick to the flowers and, even then, cook them. The trunk and branches have thorns, which can cause severe reactions if skin is scratched.

Key edible uses

The flowers add a sweet, floral flavour to dishes and can be used to make a fragrant tea and for flavouring jams and desserts. In Europe and Japan, the flowers are fried in batter. Flowers are edible raw or cooked, but no other parts of this plant should be eaten as even small amounts can cause health problems.

Red clover

SCIENTIFIC NAME

Trifolium pratense

DESCRIPTION

This low-growing, clumping perennial has oval leaflets (usually) produced in threes and marked with a paler V-shape. The tiny flowers cluster together to form a central ball of colour, which is white with light-purple tips.

DISTRIBUTION

Widespread throughout drier parts of Aotearoa New Zealand.

HABITAT

Often grow on tracks and roadsides. Likes open environments.

SEASONALITY

Available year-round with fresh growth in spring and flowers in late spring and summer, but given the right conditions it can flower longer.

FORAGING TIPS

It is best to collect leaves before the flowers appear, and to collect flowers when they have a rich colour.

CAUTIONARY NOTES/CONFUSION SPECIES

Avoid in pregnancy or if breastfeeding as the plant, like soy, contains isoflavones, which might influence oestrogen activity though there is insufficient research to conclude whether this is detrimental or beneficial. As the plant absorbs heavy metals, avoid foraging where contamination might be an issue.

SIMILAR SPECIES

The low-growing, spreading white clover *Trifolium repens* has similar leaves and a white flower; it can be consumed in the same way, though is easier to digest when cooked. There is insufficient research to indicate whether other varieties of clover are as beneficial. Note that the spotted medick, *Medicago arabica*, has similar three-lobed leaves to the clover, but the purple mark in the middle of the leaves and the tiny yellow pea-flowers indicate that this is a different species. While some people eat it, it contains saponins so is best avoided.

Key edible uses

Containing an array of protein, vitamins and minerals, and shown to have medicinal properties, the nectar-flavoured flowers can be used for infusions in vinegar, syrup, jelly and alcohol. Fresh flowers, especially with a rich purple colour, can also be added to salads. The dried flowers make good herbal teas. The leaves are also edible, the fresh spring growth eaten raw in salad, the older foliage steamed and eaten as a vegetable. The seed can be collected and grown into crispy microgreens.

Opposite The more colour, the more flavour is the rule with this flower.

Gorse

SCIENTIFIC NAME

Ulex europaeus

OTHER NAMES

Furze, whin

DESCRIPTION

A dense shrub growing up to 3m high with woody branches and green spiky foliage around and up the branches. Its small yellow flower is the typical pea shape, borne in clusters up the branches. Often grows in thickets and sometimes used for hedging.

DISTRIBUTION

Widespread throughout coastal Aotearoa New Zealand. As a garden and farm escapee, it is often found around or near areas of human settlement.

HABITAT

Likes well-drained and sandy soils with a high amount of sun exposure, but will adapt to a wide range of terrains and altitudes.

SEASONALITY

In flower all year but with a peak in flowering in spring.

FORAGING TIPS

Use gloves and have protective clothing when gathering, as the branches are covered in spikes.

CAUTIONARY NOTES

The seeds contain lectin, which can prevent the body absorbing certain nutrients, it's best to avoid eating them; focusing on the flowers will avoid this.

Although broom (*Cytisus*, *Genista* and *Spartium* spp.) does not have spikes and its foliage is different, the shrubby form, yellow pea-flowers and seedpods could be confused with gorse. It's generally fine to eat broom as a garnish, but don't consume in any great quantity.

Key edible uses

The flowers are used when bright yellow and add a pop of brightness to salads. They have an aromatic coconut flavour that can be infused into oil, vinegar and alcohol. Wine and herbal teas can be made from gorse flowers.

Opposite Gorse flowers are a striking bright yellow. **Inset** Use gloves when gathering gorse, as the branches are covered with sharp spikes.

Common vetch

SCIENTIFIC NAME
Vicia sativa

DESCRIPTION
A slender-stemmed, sprawling plant typically ranging in height from 0.25 to 1m. The small oval leaflets growing up the stalks are paired, have tiny hairs and lead to curly tendrils at the end. The small purple flowers are the typical pea-flower form producing pea-like pods, which contain pea-like seeds.

DISTRIBUTION
Found throughout Aotearoa New Zealand but absent in drier parts of Central Otago and the west coast of the South Island. Prefers low- to mid-altitude environments.

HABITAT
Common in open environments that have good sunlight, often growing among or over other wild greens.

SEASONALITY
Grows throughout the year but most prolific over spring.

Inset Vetch has ornate tendrils with tiny purple flowers. **Opposite** Fresh vetch growth with curly tendrils.

Key edible uses

With its distinct pea flavour, the foliage is tasty in salads and the fresh-growth tendrils and ornate purple flowers make a good garnish. Gather vibrant green growth as the older foliage can become quite tough and stringy, especially when foraged in drier conditions. The small seeds should be soaked, rinsed and cooked well before eating.

Geraniaceae

Herb robert

Geraniaceae is commonly known as the geranium or cranesbill family, both names referring to the long-beaked seedhead in some of its members, which supposedly looks like the head of a crane. One section of this family is the pelargoniums, also known as the storksbill family because their seedhead is supposed to resemble a stork, with a small head and long beak. To further complicate the situation, pelargoniums are often called geraniums. For the forager, both wings of the family have edible plants.

They mostly have rounded leaves, though these may look crimped or in some cases deeply divided. The flowers have five petals, but while the petals in the cranesbill flower are similar to each other, in pelargoniums the top two petals are slightly different from the bottom three. The flowers are generally in the colour range of pink, red, purple, blue and white, and some varieties may be bicoloured, have darker lines on some of the petals, or have a distinct eye in the centre. Cultivated varieties of pelargonium generally bear their flowers in tight clusters, producing balls of colour. In wild forms, though, the flowers and plants — like the cranesbill part of the family — tend to be less showy and more delicate in appearance.

There are a number of threatened indigenous geraniums, such as *Geranium solanderi,* the roots of which were part of both Māori and Australian Aboriginal diets. However, with rabbits now nibbling the foliage their number is reducing, so foragers need to bear in mind that some geraniums need to be left alone.

Geranium species

SCIENTIFIC NAME

Geranium spp.

DESCRIPTION

The geraniums you are likely to come across will be mostly low-growing and have small delicate flowers and more-dominant stems and foliage, which is likely to be rounded, toothed and 'cut-leaved', as the common name of *Geranium dissectum* (cutleaf geranium) describes it. They have five identical petals which produce the head of seeds with the characteristic beaky spike.

DISTRIBUTION

Found throughout Aotearoa New Zealand except for alpine or heavily forested zones. The cranesbills are widespread in drier parts of the country.

HABITAT

Geraniums adapt to a wide range of well-drained soils, but prefer drier, open, sunny areas, and are often found around the edges of vineyards and farmland.

SEASONALITY

Geraniums typically bloom during the spring and summer months, though some species may continue to produce flowers intermittently throughout the year.

Opposite Geranium flowers come in a range of colours and have a slightly minty flavour.

Key edible uses

Geraniums have a slightly sweet flavour. The fresh leaves are best shredded and used in salads and herb butters, or else infused to make herbal tea. Geranium seeds can be used as a herby aromatic sprinkle, which is reported to have a taste similar to rosemary. Geranium flowers are slightly minty and make delicate garnishes. The stems and roots were eaten traditionally but are bitter.

Dove's-foot cranesbill

SCIENTIFIC NAME
Geranium molle

DESCRIPTION
The downy leaves are rounded but indented and divided, supposedly resembling the shape of a dove's foot. The small flowers have the usual 5 petals, but are double-petalled so can look as though they have 10. They are typically purple, but can also be blue or whitish. Each flower has a distinctive central column with multiple stamens. The seeds have the characteristic long, pointed beak.

DISTRIBUTION
Widespread throughout Aotearoa New Zealand but absent from alpine zones and heavily forested parts of the west coast of the South Island.

HABITAT
Typically found in disturbed areas, open fields, gardens and along roadsides. It can thrive in a range of soil types but prefers well-drained areas. Common around the edges of vineyards and in open environments with good sunlight.

SEASONALITY
Available year-round with peak growth and flowering in spring.

FORAGING TIPS
Best collected after damp periods when there is new growth, as older plants become tough and bitter.

Key edible uses

The flowers can be used as a garnish. The mild-flavoured fresh leaves can be added to salads, or lightly steamed and eaten as a green. Both leaves and flowers can be dried for herbal teas.

Inset The double-petalled dove's-foot cranesbill flower is typically purple, can be blue, or white.

Herb robert

SCIENTIFIC NAME
Geranium robertianum

DESCRIPTION
A low herb growing up to 30cm high, it has deeply lobed leaves that are more triangular than round and are often tinged with red or purple. The stems also have a distinct reddish tinge. The leaves produce a pungent odour when crushed. Has small pink flowers.

DISTRIBUTION
Throughout Aotearoa New Zealand, generally in lower-altitude areas but is found in the Queenstown region. Most common in the upper half of the North Island, particularly Auckland.

HABITAT
Likes areas of damp ground and shade, favouring edge environments, close to urban centres and townships.

SEASONALITY
Available year-round with peaks in growth in spring and also a lesser flush in autumn. Flowers mainly in spring and summer.

Key edible uses

Containing vitamins B and C and various minerals, the leaves are edible and have a slightly sour taste, so are mainly used as a herbal flavouring, but they can be added to salads or used, along with the flowers, as a garnish. The leaves can also be dried to make tea. The plant was traditionally used for healing and does contain geraniin, a tannin that has medicinal properties.

Inset Herb robert has a distinctive red stem.

Scented geraniums

SCIENTIFIC NAME

Pelargonium spp.

DESCRIPTION

While there are a number of different types of *Pelargonium*, the most edible are the ones you can identify just by rubbing and sniffing their strongly aromatic leaves. There is a wide range of highly scented varieties: lemon, rose, pineapple, peppermint, ginger and many others with distinct perfumes. Usually growing as a small shrub, with hairy stems. The five-petalled flowers are borne in showy heads, the scented varieties in this country usually being pink, but sometimes white or purple. The rounded leaves often have a crinkled appearance.

DISTRIBUTION

Mostly found around Auckland, Wellington and Christchurch, but a few scatterings through much of the country.

HABITAT

Thrives in sunny, frost-free areas and in sandy, well-drained soils. Often found on the edge of gardens, around settlements and urban areas.

SEASONALITY

Leaves are available year-round, flowers in summer.

Opposite Scented geraniums make the best eating and come in an array of aromatic varieties.

Key edible uses

Providing a slightly sweet flavour, the leaves generally taste like their smell, be it lemon, rose, ginger, orange, or the many varieties you might come across. Given their thick texture, the leaves are palatable only when finely shredded but are better dried and crushed to add flavouring in baking or savoury dishes, especially crockpot cooking. Alternatively, use them to infuse teas, syrups, vinegars, cocktails, desserts, baking, etc., and then discard. The milder flowers can be eaten raw in a garnish or salad.

Lamiaceae

Wild oregano flowers

Also known as the mint family, Lamiaceae brims with nutritious and delicious culinary herbs, including basil, oregano and sages. The name refers to the Latin for deadnettle — one of its members — but the family was previously known as Labiatae, in reference to the flowers of many of its members, which have petals fused into lobes, creating a sort of upper and lower lip and resulting in a slightly squashed, irregular appearance. The flowers can be large or tiny and not all members have this type, but it can be a useful step in identification, along with other common features such as pairs of leaves going up squarish stems.

It's an aromatic family, too, rich in volatile oils such as menthol; so if your foraging has a strong herby perfume — as in rosemary, lavender or mint — Lamiaceae membership is a strong possibility.

Ground-ivy

SCIENTIFIC NAME
Glechoma hederacea

OTHER NAMES
Alehoff, cat's-foot, field balm, creeping charlie, gill-over-the-hill, hay maids, runaway robin

DESCRIPTION
A perennial herb in the mint family with creeping stems that grow into dense patches up to 20cm high. The small leaves are scalloped along the margins and have a heart shape. The tiny purple flowers grow in clusters.

DISTRIBUTION
Scattered locations throughout Aotearoa New Zealand, mainly near urban centres and generally in lower-elevation sites.

HABITAT
Can be found in both shaded and sunny locations, mainly in damp areas and often on forest margins or edge zones of gardens, particularly as groundcover under hedges.

SEASONALITY
All year round, with flowering in both spring and autumn.

CONFUSION SPECIES
Ground-ivy looks quite different, but do not confuse with ordinary ivy, *Hedera helix*, all parts of which are poisonous.

Key edible uses

The aromatic new-growth leaves can be eaten raw in salads or cooked as a side green. Dried, the leaves are a useful aromatic flavouring for all sorts of dishes and along with the dried fresh flowers make a colourful plating garnish.

Opposite Ground-ivy grows in dense patches, often in damp areas and edge zones.

Deadnettle

SCIENTIFIC NAME
Lamium purpureum

OTHER NAMES
Purple archangel, purple deadnettle, red deadnettle

DESCRIPTION
A low-growing annual plant that reaches about 20cm high. The paired leaves are heart-shaped or oval, with serrated margins, and are green or reddish-green with a purplish tinge. The square stems can also be reddish or green, and they often have fine hairs. The flowers of deadnettle are small and have the two lips, arranged in whorls in the leaf axils. They are typically reddish-purple, although the colour can vary from white to pink to purple.

DISTRIBUTION
Throughout Aotearoa New Zealand but absent from drier inland and alpine regions.

HABITAT
Found in a variety of habitats, including gardens, waste areas, roadsides, meadows and disturbed soils. It thrives in both sunny and shaded locations but generally likes damp soil.

SEASONALITY
Fresh growth appears in early winter, with flowering from late winter to early autumn when it dies back.

SIMILAR SPECIES
Aluminium plant (*Lamium galeobdolon*) also known as golden deadnettle or yellow archangel can be eaten in a similar way. It has whorls of yellow flowers and its nettle leaves have white mottling over the veins.

Key edible uses

The young leaves are edible and can be consumed raw in salads or cooked as a side green. They have a mild, slightly earthy flavour. While dead nettle is not as well known as other wild edible plants, it has been historically used as a leafy green in some regions. The flowers can also be used as a garnish or dehydrated with the leaf and added into a herbal tea mix.

Lemon balm

SCIENTIFIC NAME
Melissa officinalis

DESCRIPTION
A perennial herbaceous plant growing up to 1m high and often in small thickets. The bright-green leaves are pointed ovals with a crinkled-looking surface and toothed margins, and when crushed have an aromatic lemon smell. It has very small, white, lipped flowers growing on the leaf axils.

DISTRIBUTION
Found throughout Aotearoa New Zealand, often in areas near habitation but occasionally in abandoned edge zones, farmland and river margins.

HABITAT
Likes fertile soils with moderate moisture. Often occurs in sheltered and shaded sites.

SEASONALITY
Fresh growth in spring and flowers through summer. Tends to die back over winter.

Key edible uses

The leaves are used as a lemon-flavoured aromatic seasoning for a range of dishes — sweet or savoury — and the fresh growth can be added to salads. The leaves — fresh or dried — make excellent herbal teas.

Wild spearmint

SCIENTIFIC NAME

Mentha spicata

OTHER NAME

Mint

DESCRIPTION

Typically grows in tight patches up to 30–70cm high with spreading branches on square-shaped stems. The leaves are bright green, spear-shaped, with serrated edges and grow in pairs, emitting a pleasant fragrance when crushed or bruised. Spearmint produces very small, delicate flowers that cluster together in spikes at the top of the stems and are typically pale to dark purple.

DISTRIBUTION

Throughout Aotearoa New Zealand but prefers areas with high amounts of sunshine. It is common around cities and at higher altitudes, having spread through the South Island's high country. Largely absent from the west coast of the South Island.

HABITAT

Growing on the edges of freshwater streams, it is a resilient herb that can thrive in a variety of soils and habitats, particularly with moist soils and ample sunlight. It prefers fertile, well-draining soils that are rich in organic matter.

SEASONALITY

Fresh growth in spring and flowers in late spring and summer; it tends to die away in late autumn and winter.

Key edible uses

The leaves have an intense, refreshing spearmint flavour and can be used as a garnish, to flavour pesto, to make mint sauce to accompany lamb, or as a culinary herb for sweet and savoury dishes. Both the leaves and the purple flowers are a popular cocktail garnish, flavouring mojitos and mint juleps, and the leaves are especially popular for mint tea.

Apple mint

SCIENTIFIC NAME

Mentha suaveolens

DESCRIPTION

Grows from 30 to 70cm high. Its leaves are light green, oval with rounded rather than pointed ends, and a slightly furry and wrinkled texture. The spired flowerheads form on the ends of the stalks and are white to pinkish-white clusters of tiny flowers. The plant has a fruity-peppery aroma when crushed.

DISTRIBUTION

Found throughout Aotearoa New Zealand, mainly close to urban centres in coastal and lowland areas.

HABITAT

Found in areas with damp soils, often spreading in the rank vegetation along watercourses.

SEASONALITY

Available throughout the year, though there is some dieback in late winter with most of the fresh growth in spring and flowering over summer.

FORAGING TIPS

Use a sharp knife to cut the fibrous stems and then bunch them together to dry out.

Key edible uses

Leaves are used for flavouring jellies, as a garnish, added to salads, grains or couscous, or used to make mint sauce for lamb. Dried, the leaves make mint tea; fresh, they can also be fried until crisp and used to garnish fish dishes.

Oregano

SCIENTIFIC NAME
Origanum vulgare

OTHER NAME
Wild marjoram

DESCRIPTION
A bushy herb that typically grows to a height of 20–80cm. It has square-shaped stems that are woody at the base, and its leaves are pointed ovals, paired and covered with very fine hairs. The leaves have a strong aroma and a characteristic flavour. The flowerhead is comprised of clusters of tiny flowers, which are pink-purple.

DISTRIBUTION
A patchy distribution throughout many drier parts of the country, with strongholds in Central Otago and the mid to lower North Island.

HABITAT
Areas with well-drained soils and strong sunlight. Often found on the edges of tracks in higher-altitude areas.

SEASONALITY
Main flowering season is November to May.

FORAGING TIPS
Take care when foraging, as bees are attracted to the flowers.

Opposite Wild oregano is a feature in the summer, cropping up along mountainsides in Central Otago.

Key edible uses

A very useful herb, used in many Mediterranean dishes such as pizza, it goes well with strongly flavoured meats but also in a range of vegetarian dishes. Leaves and flowers can be dried to make herbal tea. The leaves keep well when dried.

Self-heal

SCIENTIFIC NAME

Prunella vulgaris

OTHER NAMES

Brownwort, carpenter's herb, heal-all, heart-of-the-earth, woundwort

DESCRIPTION

A small species growing to 20cm high. The lance-shaped leaves have serrated edges, are slightly hairy and grow up the stems in pairs. It has small, tubular flowers that cluster together in dense, cylindrical spikes at the top of the stem. The flowers are two-lipped and can range in colour from shades of purple, blue and pink to occasionally white.

DISTRIBUTION

Widespread throughout Aotearoa New Zealand except in heavily forested areas and alpine zones.

HABITAT

Grows in a wide range of edge or open environments, areas with well-drained soils and good sunlight.

SEASONALITY

Available throughout the year but peak growth from October to May.

Inset The purple self-heal flower has two lips.

Key edible uses

The whole plant is edible, but fresh growth has the best flavour for salads and as a garnish. Mature plants can be used as a herb added to soups and stews. When infused fresh or dried, the leaves impart a refreshing flavour to drinks, and the dried leaves can be used for herbal tea. The flower is useful for infusions to flavour jellies and vinegars. While not toxic, the leaves and flowers are bitter and have a strong medicinal action. Young micro-greens can be soaked to reduce bitterness. Otherwise, use sparingly, as a garnish rather than a vegetable.

Hedge woundwort

SCIENTIFIC NAME
Stachys sylvatica

OTHER NAMES
Hedge nettle, hedge woundwort, whitespot

DESCRIPTION
Growing to 80cm high from a single stalk, this perennial has toothed oval leaves with a pointed tip. The flowerheads are spires comprising tiny purple flowers at the tip of each stalk. It has a slightly unpleasant smell when fresh.

DISTRIBUTION
Found throughout Aotearoa New Zealand; tends to be more common in the North Island in areas of low to mid elevation.

HABITAT
Likes areas of damp ground in edge environments, close to shrub cover.

SEASONALITY
Available year-round, flowering from mid-spring to early summer. Fresh-growth leaves are available in late winter and mid-autumn.

Key edible uses

The fresh-growth leaves are edible when steamed, and can be used as an aromatic herb for flavouring a range of dishes.

Wild thyme

SCIENTIFIC NAME

Thymus vulgaris

OTHER NAMES

Common thyme, garden thyme, German thyme

DESCRIPTION

A small, low-growing herbaceous shrub that often grows in patches. The small narrow leaves are only 2–4mm long, grey-green with a hard texture. The tiny flowers are either white or purple and arranged in clusters at the top of the stem. Thyme has a strong earthy-mint aroma.

DISTRIBUTION

Central Otago is the stronghold for wild thyme, especially around Lake Dunstan in the Kawarau Valley, but there are patches up and down the country, particularly around cities.

HABITAT

Well-drained soils, often on hillsides that are exposed to strong sunlight.

SEASONALITY

Available year-round but fresh growth in late winter and spring provides the best flavour, with flowering in mid to late spring.

FORAGING TIPS

The whole plant can be easily removed and grown in a pot or planted in the garden.

Key edible uses

The flowers and leaves have a strong aromatic flavour and can be added fresh to salads. The fresh or dried leaves can be used for herbal teas or as a seasoning for a wide range of dishes, for example sprigs being added to slow-cooked meals. The flowers can be used as a flavour infusion in vinegar. The wood can also be used for lightly smoking foods.

Inset and opposite Wild thyme grows well on drained hillsides, and can be found easily in Central Otago.

Malvaceae

Tree mallow

This is a particularly varied botanical order, ranging from small plants to tall trees — including the cacao tree, where chocolate comes from. *Hibiscus* species are also part of this order, and if you bear in mind how the standard hibiscus species have palm-shaped leaves and showy five-petalled flowers holding a prominent stalk of fused stamens in the centre, then you'll recognise some of the identifying features that are commonly found in Malvaceae. Although not included in this listing (as it is more a garden plant than a wilding), hibiscus is edible: the flowers make a tangy tea and the nutritious seed can be a coffee substitute. Like many other Malvaceae members, the hibiscus calyx (or sepal-enclosed flower bud) and the leaves are also edible. Again typical of the family, in some varieties the leaves are slightly hairy and contain a mucilaginous substance, which gives a gelatinous texture and can thicken dishes such as soups and stews. Shredding or boiling the leaves makes their hairy and gummy texture more palatable.

The most famous use of this mucilaginous property is in marshmallow, once made from the root of the marsh mallow (*Althaea officinalis*) plant. While there are some wildings of marsh mallow in the north of the west coast of the South Island, with some spread in Waikato, it is too rare to feature in these pages. However, there are other mallows that are readily available for foraging in Aotearoa New Zealand. Indeed, they have a tendency to become invasive; so although they can look spectacular, don't worry about harvesting them. However, as the leaves absorb nitrates from the soil, do not forage areas high in nitrogen (for example where there's runoff from highly fertilised or manured gardens and farmland, or near wastewater treatment plants or industrial zones).

There are many more edible Malvaceae than listed in this section, and new species are becoming naturalised in Aotearoa New Zealand. For instance, the flowers of abutilons are edible. There are many more malvas spreading than detailed below, and as they are high in vitamins, particularly vitamin A, as well as various beneficial minerals, it is also worth keeping an eye out for others — such as the Cretan mallow, *Malva multiflora*; the musk mallow, *Malva moschata*; the bull mallow, *Malva nicaeensis*; and the small-flowered mallow, *Malva parviflora*. Meanwhile, hollyhocks, *Alcea rosea*, are also edible, though the hairs on their leaves and stems can cause irritation. As always, careful identification is essential when you encounter an unfamiliar plant in your foraging.

Tree mallow

SCIENTIFIC NAME
Malva arborea

OTHER NAMES
Lavatera arborea, sea lavatera, sea mallow, velvet leaf, wild tree mallow

DESCRIPTION
A hardy, fast-growing shrub that is usually about 2m high but can reach up to 3m and spreads out to 2m in width. The lobed leaves are wide, up to 20cm, and have a slightly furry texture. The deep-pink-purple funnel-shaped flowers can be as large as 4cm wide and have darker-purple nectar guides and centre. They produce green pods in clusters, which turn brown when the seeds ripen.

DISTRIBUTION
Found throughout Aotearoa New Zealand but tends to be more common in drier east coast regions and is largely absent from high rainfall areas in Te Tai Poutini West Coast.

HABITAT
Likes well-drained soils generally in areas of low rainfall. Most common along coastal regions, preferring some shelter so often found close to rock walls. Grows in open and well-lit environments.

SEASONALITY
Available, even in flower, throughout the year, but the foliage can be tough and less flavourful after long dry periods.

FORAGING TIPS
Avoid foraging when leaves have any mottled discolouration.

Key edible uses

As they are hairy, the large green leaves are used as a serving plate or disposable food wrap. Younger leaves are good shredded in a salad. The green calyx buds have a crunchy texture and more flavour than the leaves, and can be pickled to make caper alternatives or eaten fresh in a salad. The purple flowers are very attractive as a garnish. Overall, the tree mallow has a very mild flavour. The firm green seedpods have a crunchy texture and slightly more flavour than the rest of the plant.

Opposite Tree mallow in flower.

Island mallow

SCIENTIFIC NAME
Malva assurgentiflora

OTHER NAMES
Lavatera assurgentiflora, Californian mallow, malva rosa, mission mallow, royal mallow

DESCRIPTION
Grows to about 3m high with a central woody stem and bushy habit. The lobed leaves are very large, grey-green and with more-pointed edges than other mallow species. Has rich pink flowers with darker nectar lines.

DISTRIBUTION
Found wild on the east coast of the South Island from Blenheim to Dunedin, with a very patchy distribution often near coastal settlements.

HABITAT
Dry, well-drained coastal locations, often growing right on or near the coastline.

SEASONALITY
Foliage and flowers available throughout the year.

Key edible uses

The leaves, calyx buds and flowers are all edible, with the flowers having a mild sweet taste. The leaf and flower buds are firm and crunchy and can be pickled.

Inset Mallow flowers have a deep pink or purple colour with distinct highlights.

Common mallow

SCIENTIFIC NAME
Malva sylvestris

OTHER NAMES
Cheeses, French hollyhock, high mallow, large-flowered mallow, tall mallow

DESCRIPTION
Common mallow is a herbaceous perennial plant that typically grows to a height of 30–80cm. It has round to kidney-shaped leaves that are often shallowly lobed or toothed. The leaves are soft and hairy, giving them a velvety texture. The flowers are usually deep pink or purple, although white-flowered varieties can also be found. The flowers have five petals, darker nectar guides and are approximately 2–3cm across. In dry conditions it can grow as a mat close to the ground. Lacks the woody stem of the tree mallow (see page 176).

DISTRIBUTION
Found throughout Aotearoa New Zealand in coastal and mid-elevation areas, but largely absent from higher altitudes above 800m and the west coast of the South Island.

HABITAT
Grows in a wide range of areas, favouring dry soils and open environments with good amounts of sunlight.

SEASONALITY
Available throughout the year but tends to die away over the winter months and during periods of frost.

Key edible uses

Mildly flavoured and high in vitamins and minerals, the new leaves are consumed raw or cooked, though are particularly good shredded raw into salads. Although they become gummy when cooked, the leaves can be served as a side green, boiled or fried as a substitute for spinach leaves, or added to soups. Being edible, the raw flowers are useful for garnishes. The round seedpods, which appear soon after flowering, were once referred to as cheeses and eaten fresh; they add crunch when sprinkled onto salad. The seeds make a nutty sprinkle, while a herbal tea can be made from the dried leaves, flowers and/or seeds.

Dwarf mallow

SCIENTIFIC NAME

Malva neglecta

OTHER NAMES

Buttonweed, cheeseplant, cheeseweed, common mallow, roundleaf mallow

DESCRIPTION

A mallow with small, rounder leaves that grows in a prostrate manner, creeping along the ground. Typically has small white flowers with deep-pink nectar guides, though the petals can be a very light pink.

DISTRIBUTION

Widespread throughout Aotearoa New Zealand, especially in drier regions in the South Island.

HABITAT

Likes areas of dry open ground and is often a coloniser species on disturbed or bare soil.

SEASONALITY

Available throughout the year, but tends to die back over the winter months with fresh growth in spring to autumn.

FORAGING TIPS

Best to harvest after rain when there is fresh growth.

Key edible uses

Flowers, stems, calyx buds and leaves are all edible and mildly flavoured like the other mallow species, and can be eaten raw or cooked. The leaves are good raw in salads; when cooked they have a gummy texture but can be treated as a side vegetable or else dried to make tea. Steeping the chopped leaves in water for a couple of hours will also extract a quantity of the mucilaginous substance that can be used as a thickener, salad dressing or a vegan substitute for egg white. The thickened water from peeled, sliced and boiled roots also works as a substitute for egg white, though it may be tricky to make meringues with them. The slightly nutty seeds are high in protein; when ripe, cook them as a substitute for grain.

Linden tree

SCIENTIFIC NAME

Tilia spp.

OTHER NAMES

Lime, tilia

DESCRIPTION

A relatively recent reclassification has made the linden tree a member of the Malvaceae family. There are various *Tilia* species in New Zealand, which are medium to large-sized trees, typically between 18m and 25m high, with a pyramid-shaped or rounded crown with dense foliage. The leaves are heart-shaped with serrated edges. The bark is usually greyish-brown and develops grooves with age. The tree produces tiny sweetly fragrant flowers that are pale yellow and typically arranged in clusters hanging beneath the branches.

DISTRIBUTION

Found throughout the country, but with a patchy distribution mainly around urban centres and also in inland Central Otago.

HABITAT

Often found in parkland areas and plantings around established homesteads.

SEASONALITY

Available all year, with flowers in late spring to early summer.

Key edible uses

Mildly flavoured, the new leaves can be foraged immediately after bud burst and eaten raw in salads. The young shoots can be lightly steamed. The flowers contain antioxidants; when fresh they make a tasty garnish and when dried a sweetly flavoured, aromatic herbal tea, which is not surprising given that linden honey is particularly prized. The inner bark from the tree can also be used to make a flour.

Inset The leaf of the linden tree, with seed clusters forming.

Myrtaceae

Mānuka in flower

183

Also known as the myrtle family, the members of Myrtaceae are generally woody shrubs or trees, produce essential oils, and their flowers often have an impressive number of stamens which can give them a fluffy appearance. The oil of some species can be potent, making them inedible or risky. A few dried eucalyptus leaves can be used to infuse tea or savoury and sweet dishes, but it contains eucalyptol, which is toxic in high doses. Myrtaceae, however, is a family that you can incorporate into your cooking in fresh ways.

The Antipodes have been blessed by some myrtle members all their own, and the following are iconic trees of either Aotearoa New Zealand or Australia. Currently these are found in abundance in this country, but the arrival of the plant disease myrtle rust may deplete their numbers; if so, we may have to remove them from our foraging lists.

Kānuka and mānuka

SCIENTIFIC NAME
Kunzea ericoides, Leptospermum scoparium

OTHER NAMES
Tea tree, white tea tree

DESCRIPTION
Both are dense, dark-green shrubby trees; mānuka growing from 2 to 8m and kānuka up to 15m. Both produce white five-petalled flowers (though varieties of mānuka come in pink and red). Kānuka has short lance-like leaves, which look spiky but are soft, while those of mānuka are tough and prickly. Both have an aromatic smell when rubbed. The seeds of mānuka are notably larger than those of kānuka.

DISTRIBUTION
Mānuka is one of the most widespread shrubs in Aotearoa New Zealand from coastal to alpine zones, although naturally largely absent from the Canterbury Plains. Kānuka grows across much of the country but is rarer in the far north and south.

HABITAT
Found in scrub and edge areas, especially native forests and waterways on both flat ground and hillsides. Kānuka is a coloniser species after forest clearance, forming dense thickets.

SEASONALITY
Fresh-growth foliage is available throughout the year, but mostly in late winter and early spring. Flowers from October to February.

FORAGING TIPS
Take a hand-saw to harvest and use the sawdust for smoking foods. Young growth has a milder flavour and often grows low so can be easily hand-harvested.

Key edible uses

The wood has a high burning point and produces an aromatic smoke that can be used to cold- and hot-smoke a wide range of foods. The twigs can be used for flavour infusions, doing double service when used as skewers. A refreshing herbal tea can be made from sprigs of fresh green growth, and the flowers, through simply steeping the branches in hot water, will also create a refreshing drink. The flavour is similar, though kānuka is sweeter. The seeds can be roasted and ground to be used in a spice mix. The plant has strong antibacterial and anti-inflammatory properties as well as other beneficial qualities.

Clockwise from left Mānuka flowers; mānuka seeds; kānuka in flower; fresh kānuka foliage.

Pōhutukawa

SCIENTIFIC NAME
Metrosideros excelsa

OTHER NAME
New Zealand Christmas tree

DESCRIPTION
Typically grows to a height of 10 to 20m, often with multiple trunks and creating a wide-spreading canopy. The dense, dark-green leaves are leathery ovals. The trees produce clusters of vibrant crimson flowers that have a tufted appearance as they are made up of a huge number of stamens, sometimes tipped with yellow, though there are yellow and orange varieties. The flowers are about 2–3cm in diameter.

DISTRIBUTION
Widespread from the upper half of the South Island northwards. Ornamental plantings have extended the species' distribution.

HABITAT
Thrives in frost-free coastal environments with well-drained soils, often along coastal margins, including on cliffs, rocky outcrops and sandy beaches.

SEASONALITY
Blooms from November to February.

FORAGING TIPS
Harvesting the stamens is time-consuming and needs to be done by hand. The best flavour is obtained when the stamens also have pollen on them.

Key edible uses

The stamens can be used to flavour fizzy drinks after being boiled and left to cool, resulting in a red-coloured liquid with a sweet, floral, lemony flavour.

Above inset Pōhutukawa flowers are typically scarlet red, but there is also a yellow cultivated variety which is a less common sight around the country. **Opposite** Yellow pōhutukawa often occurs around urban areas

Lilly pilly

SCIENTIFIC NAME

Syzygium spp.

OTHER NAMES

Acmena spp., monkey apple

DESCRIPTION

Lilly pilly plants can vary in size from small shrubs to tall trees, depending on the specific variety and growing conditions. Some varieties are commonly grown as hedges or ornamental shrubs. The leaves are glossy, elliptical to lance-shaped, and typically dark green. They may have a wavy margin and are often used as a screening or decorative feature in landscaping. Lilly pilly produces clusters of small fluffy flowers that are cream-coloured to white. The fruit is typically slightly smaller than cherries, rounded, hard and berry-like. It can range in colour from white to pink to purple, depending on the variety and maturity.

DISTRIBUTION

Found from Christchurch north, but especially common on Kāpiti Coast and the upper half of the North Island, notably in the Waikato and Auckland regions.

HABITAT

Grows in areas of good rainfall; will infiltrate the bush as it likes damp and shaded environments.

SEASONALITY

Fruits mainly in summer and autumn, but trees sometimes produce small amounts of fruit through the year.

CAUTIONARY NOTES/CONFUSION SPECIES

The berries of lilly pilly may seem an easy way to identify the plant, but there are toxic trees that also dangle such fruit. Ngaio (*Myoporum laetum*) produces white, pink or purple berries, which along with the leaves contain a toxin that can cause liver damage. You can identify a ngaio tree by the tiny yellow or white dots across its dark-green leaves. The pink berries produced by the spindle tree (*Euonymus europaeus*) and Japanese spindleberry (*Euonymus japonicus*) may at first sight look like those on a lilly pilly. However, beneath the pink layer of the *Euonymus* berries is a bright-orange coating around its seeds. Both berry and seeds are toxic, so check for that orange layer.

Key edible uses

The fruits of lilly pilly are edible and have a sweet to tart flavour. They can be eaten fresh or used in a variety of culinary applications, including juices, jams, jellies, pies, syrups, chutneys, sauces and desserts. The fruit is rich in vitamin C and antioxidants.

Opposite Ripe lilly pilly fruit is usually a light purple colour that grow in clusters.

Oxalidaceae

Dock seeds

In Aotearoa New Zealand we have about 25 species of *Oxalis*, or wood sorrels, from the Oxalidaceae family. Globally there are over 2500 species. In some ways the name 'sorrel' is misleading as it is also used for the dock family (Polygonaceae, see page 222). All species of wood sorrel have ornamental flowers, the leaves being clover-like in general appearance. Some species, such as the shamrock wood sorrel, can grow in dense clumps; other species, such as the pale-pink wood sorrel, have a sprawling nature. Some, such as *Oxalis triangularis*, are grown as ornamentals and the wild species can easily be potted and grown in a kitchen garden (even indoors). There is a small woody tuber at the base of the plant (which is not practical to eat). Wood sorrels need a frost-free environment, and for that reason are absent from many inland areas.

All of the species are edible, with the leaves and the flowers being the main part eaten. In recent years, given the ornamental nature of the leaves and flowers, wood sorrel has become popular with chefs as a garnish. The sharp tangy flavour complements a range of savoury and seafood dishes, while its citrus tang balances out sweetness when sprinkled over desserts. The stems can also be used if juiced.

Wood sorrels contain the toxin oxalic acid (see page 12), though fermented food and probiotics will break that down, and blanching and discarding the water will reduce the levels, too. Thankfully, it would be very difficult to eat an excessive amount — you would need to eat several salad bags full to encounter any problems. Sorrels are also so small that it is unlikely you would consume an excessive amount by accident as you would need to spend days to collect a dangerous amount! The leaves contain various vitamins, plus iron and other useful nutrients, and have often been used medicinally both internally and externally.

At a practical level, there are six main wood sorrel species that can be foraged in this country, being of a good size, abundance and flavour. However, given the roughly 25 species that have made Aotearoa New Zealand home, you could encounter many others — such as the yellow-flowered *Oxalis chnoodes*, spreading from Auckland northwards, or the more widely scattered yellow-flowered *Oxalis rubens*, or the pink *Oxalis latifolia*, commonly known as fishtail oxalis because of the shape of its leaves. There is also the white-flowered *Oxalis magellanica*, or tūtae kāhu, which is curious because it is indigenous to this country but also to Tasmania and parts of South America.

Shamrock wood sorrel

SCIENTIFIC NAME
Oxalis articulata

OTHER NAMES
Pink sorrel, sourgrass

DESCRIPTION
A medium-sized wood sorrel with three green leaves and a small pink flower. Often grows in tight clumps, allowing for easy harvesting of large amounts of foliage.

DISTRIBUTION
Throughout Aotearoa New Zealand, preferring drier lowland regions.

HABITAT
Likes open frost-free environments with well-drained soils and good sunlight. Often found on farmland and coastal margins.

SEASONALITY
Peaks in autumn and spring, dying away in the late winter and mid-summer months.

FORAGING TIPS
Grows in tight clumps, so a large number of leaves can easily be foraged. A good species to collect from the wild and transplant into a garden to extend availability. Can be potted and grown indoors.

CAUTIONARY NOTES
Often develops an unsightly orange rust under leaves later in the year, but still edible in small quantities.

Key edible uses

One of the main *Oxalis* species used in Aotearoa and a favoured species for local chefs to use as a garnish. The long stems can also be collected in bunches for juicing, which can then add an intense, sour, lemony flavour to a wide range of dishes. The medium-sized green leaves and bright-pink flowers make a top garnish and can be added to a salad and eaten raw.

Above inset Fresh growth leaves. **Opposite** Shamrock wood sorrel is often noticed when in flower and often grows in clumps especially along coastal edge zones.

Creeping wood sorrel

SCIENTIFIC NAME
Oxalis corniculata

OTHER NAMES
Sleeping beauty, yellow sorrel

DESCRIPTION
With small crimson-brown leaves and yellow flowers, this sorrel tends to grow in creeping mats. One of the smallest species of edible wood sorrel.

DISTRIBUTION
Common in urban and suburban environments. More widespread and frost-hardy than other wood sorrel species.

HABITAT
Grows on gravel and often from cracks in concrete. Often found in and around cultivated plants.

SEASONALITY
Available year-round, but does die away in winter, especially after hard frosts.

FORAGING TIPS
As it grows close to the ground, wash thoroughly to remove any grit.

CAUTIONARY NOTES
As with all *Oxalis* species, eat in moderation.

Key edible uses

Leaves make an ornate garnish and again have the characteristic citric flavour. The leaves will add a citric bite to salads and cooked greens. Rich in vitamin C and various beneficial minerals, you can infuse the leaves in hot water, add honey and drink as a tea, either hot or chilled.

Least yellow sorrel

SCIENTIFIC NAME
Oxalis exilis

OTHER NAMES
Creeping oxalis, shady wood sorrel

DESCRIPTION
A small, creeping sorrel that grows on long stalks in clumps and has a bright-yellow flower. Its leaves are made up of three heart-shaped sections and range from green to purple. Unlike many other sorrels this species does not have bulbils, the tiny secondary bulbs.

DISTRIBUTION
Widespread throughout the country, though in patchy distribution.

HABITAT
Very common in and around gardens in urban environments, preferring sunny locations, often growing in lawns.

SEASONALITY
Available throughout the year.

FORAGING TIPS
Very delicate in structure, so handle with care and put in containers to avoid crushing. Pick flowers when sun is shining and they are fully open.

Inset Least yellow sorrel has a disproportionately large yellow flower.

Key edible uses

The leaves are nutritious and bring a sour tang to salads when used raw, or when cooked with other vegetables or added to stews and soups. The raw flowers are useful as a garnish. Contains oxalic acid, see page 12.

Pale-pink wood sorrel

SCIENTIFIC NAME
Oxalis incarnata

OTHER NAMES
Lilac oxalis, pale-pink oxalis

DESCRIPTION
A small-leafed wood sorrel that grows on long stalks in loose spindly clumps and has a proportionally large whitish-pink flower. Leaves are heart-shaped and light green in colour.

DISTRIBUTION
Widespread and common in the upper half of the South Island and throughout the North Island, but also found around Dunedin. Generally grows in lowland and coastal locations.

HABITAT
Often found in and around gardens in urban environments, preferring the damp and shady microclimates in sheltered, frost-free locations, often around the base of trees.

SEASONALITY
Available throughout the year, but peak growth is in spring with a secondary peak in autumn.

FORAGING TIPS
Very delicate in structure, so handle with care and put in containers to avoid crushing. Pick flowers when sun is shining and they are fully open.

CAUTIONARY NOTES
As it has a sprawling nature, carefully check when harvesting that there are no unwanted plants accidentally collected, as they may be toxic. Eat small amounts only.

Key edible uses

With their mild, sour flavour, the flowers and leaves can be used as a garnish.

Inset and opposite Pale-pink wood sorrel flowers year round. It has a slightly off-white flower with a pink tinge.

Bermuda buttercup

SCIENTIFIC NAME

Oxalis pes-caprae

DESCRIPTION

A long-stemmed wood sorrel with distinct brown speckling on leaves and a yellow flower, giving the plant its name. Often grows in large beds, and when flowers are open it has a strikingly vivid-yellow appearance.

DISTRIBUTION

Its strongholds are Wellington, especially Kāpiti Coast, and around Auckland and Nelson, but can be found scattered at a wide range of sites in the North Island and on the east coast of the South Island as far south as coastal Christchurch.

HABITAT

Generally prefers coastal areas with well-drained soils. Tends to grow in dense patches, liking quite open coastal environments and often found in areas with sandy soils.

SEASONALITY

Peak growth is in winter and spring. It dies away over the summer months and is largely absent in early autumn.

FORAGING TIPS

A practical wood sorrel to forage in large amounts, but best placed in a container after harvesting as it is easily crushed. As it is regarded as an invasive weed, you can harvest it without any restrictions.

CAUTIONARY NOTES

Do not be fooled by the name into thinking all buttercups are edible: the common buttercup (also known as the creeping buttercup), which is part of the Ranunculus family so is no relation to sorrels, is toxic to animals and humans.

Key edible uses

A very useful wood sorrel with a good flavour, soft texture and high moisture content, making it a good addition to salads and for juicing (the stems in particular). Has good-sized, tasty, vibrant yellow flowers, which are popular for garnishes. Contains oxalic acid, see page 12.

Inset Bermuda buttercup flowers profusely in the spring. **Above opposite** The yellow flowers grow in a cluster **Below** Bermuda buttercup leaves often have a fine brown speckling.

Purple wood sorrel

SCIENTIFIC NAME

Oxalis purpurea

OTHER NAME

Large-flowered oxalis

DESCRIPTION

The largest species of wood sorrel in the country, with large rounded leaves and large pink flowers, the leaves being proportionally much smaller than the flower. Unlike the shamrock wood sorrel (see page 192), it has a yellow centre to its pink petals. Grows low to the ground in loose, sprawling mats.

DISTRIBUTION

Only found in the North Island, particularly in the Wellington region, Bay of Plenty and through into Northland, with strongholds around Auckland.

HABITAT

Usually found in open spaces, but can be found on the edge of paths. Prefers sandy soils and growing amid low groundcover, such as lawns.

SEASONALITY

Peak growth in spring when it flowers. Like many wood sorrels, it dies away over the summer months and slowly grows again through autumn and winter.

FORAGING TIPS

It is best to collect flowers on a sunny day when fully open. Can be cultivated indoors as an ornamental plant.

CAUTIONARY NOTES

As it often grows in lawn environments, check the soil quality of the collection sites, and eat small quantities only.

Key edible uses

It is edible and has a very attractive flower, but its flavour is not as good as the smaller wood sorrels, and its leaves have a slightly coarser texture. When fully open, the large flower makes a striking garnish.

Passifloraceae

Banana passionfruit flowers

You can buy various species of passionflower in garden centres, such as *Passiflora edulis* or *Passiflora ligularis*, but the varieties that you can forage in the wild are our focus here. While more species will be scrambling out of gardens, the ones that are currently the best to forage include a native species and several others that have turned into rampant pests, choking canopies and blocking out light. Among the latter is *Passiflora caerulea,* which has an attractive flower with purply-blue filaments that are white at the eye where they radiate out from; however, its oval orange fruit is not particularly tasty and its leaves contain cyanogenic glucosides (commonly mistaken for hydrogen cyanide). Cyanogenic glucosides can convert to cyanide under special conditions, but it's not likely. You can actually use the leaves in teas, as any cyanide that is potentially created is destroyed by heat. It's probably best to avoid the leaf though, and choose one of the following varieties to forage.

Black passionfruit

SCIENTIFIC NAME
Passiflora edulis f. *edulis*

OTHER NAME
Common passionfruit

DESCRIPTION
A vigorous climber, with large three-lobed leaves. Its flowers are white, with wavy white filaments turning purple at the centre and round, dark-purple fruit with orange pulp.

DISTRIBUTION
This has mostly escaped gardens only from Waikato northwards but it can also be found in wasteland in Wellington and Christchurch.

HABITAT
Scrambles up trees and banks.

SEASONALITY
Flowers from winter to the end of summer.

FORAGING TIPS
To access higher fruit, a picking pole or ladder is a major advantage.

Above inset The distinct lobed leaf of a passionfruit vine. **Below** Black passionfruit has a purple and white flower with wavy hair-like filaments.

Key edible uses

A wide range of uses in desserts and drinks or in savoury dressings. Flowers and stems are used in calming teas.

Pink-flowered passionfruit

SCIENTIFIC NAME

Passiflora pinnatistipula

OTHER NAMES

Gulupa, poro poro (not to be confused with the native shrub poroporo, *Solanum* spp.), yellow passionfruit

DESCRIPTION

This vigorous vine grows several metres high and has large three-lobed leaves and pink star-shaped flowers with purple filaments and a long stigma. Its rounded fruit starts out grey-green, ripening to yellow.

DISTRIBUTION

Spreading along the east coast of the South Island from Christchurch southwards.

HABITAT

Grows up trees and banks on scrubland and forest margins.

SEASONALITY

Flowers throughout the year.

FORAGING TIPS

To access higher fruit, a picking pole or ladder is extremely useful.

Key edible uses

An Andean delicacy, the mildly flavoured sweet-acid pulp can be scooped out and eaten raw or mixed into dressings, jams, drinks or desserts.

Inset Pink-flowered passionfruit has a picturesque, ornate flower.

Banana passionfruit

SCIENTIFIC NAME

Passiflora tarminiana, Passiflora tripartita

DESCRIPTION

Two different species but both similar and commonly named after the banana.The foliage is similar to the other passionfruit varieties with three lobes, but the flowers are larger, bright pink and face downwards. The vigorous *Passiflora tripartita*, however, has two stipules at the base of each leaf stalk, which look like rounded leaflets or if dropped off will leave irregular markings rather than a smooth clean joint. The ripe fruit of both is bright yellow and an elongated oval, ranging from 5 to 12cm in length. Banana passionfruit taxonomy is in flux, and Aotearoa New Zealand has several species.

DISTRIBUTION

Throughout frost-free coastal areas of Aotearoa New Zealand, but absent from cooler parts of the South Island. Generally found below 100m elevation.

HABITAT

Mainly in shrub and forest areas in coastal and lowland zones. Grows as a vine over a wide range of small to medium-sized trees and can form thick vines. Especially common growing over scrub on coastal hillsides.

SEASONALITY

Passiflora tarminiana fruit are available throughout the year but there is a peak over autumn (March–May). *Passiflora tripartita* flowers from December to January, with fruit through late summer and autumn.

FORAGING TIPS

To access higher fruit, use a picking pole or ladder.

Key edible uses

The sweet aromatic fruit can be eaten raw when ripe. The orange-coloured juice can be separated from the seeds and used in drinks or syrups, while the seeds can be crushed to make a form of dukkah or to provide a crunchy texture to desserts.

Inset The elongated banana passionfruit is delicious and sweet when ripe. **Opposite** Banana passionfruit vines have arresting pink flowers.

Kōhia

SCIENTIFIC NAME
Passiflora tetrandra

OTHER NAMES
Kāhia, kaimanu, kohe, kūpapa, pōwhiwhi, native passionfruit

DESCRIPTION
This Aotearoa New Zealand native looks quite different from its South American cousins, having very small oval leaves, which end in a point, and even smaller flowers and fruit. A number of flowers with white or yellow filaments appear on short stalks up the stem. The ripe fruit are bright orange in colour and only 1.5–2cm in diameter, making it easy to distinguish kōhia from the introduced passionfruit species. The ripe fruit darkens as it ages, looking like a large Jaffa.

DISTRIBUTION
Found throughout the North Island and northern South Island. Occurs as far south as Banks Peninsula.

HABITAT
Grows as a vine over lowland native forests. Generally found on the edges of native forest patches or growing over medium-sized trees, but usually absent from the canopy environment of primary trees.

SEASONALITY
Fruit are ripe from February to May.

FORAGING TIPS
To access higher fruit, a picking pole or ladder is a major advantage.

Key edible uses

The fruit generally have a low moisture content and a mild cress-like flavour, which gives the native passionfruit a distinctiveness in our culinary spectrum. The peppery seeds can be ground to make a form of dukkah.

Inset Kōhia fruit are much smaller than the introduced species of passionfruit, and are stunningly red when ripe. **Opposite** Kōhia in various stages of ripeness.

Plantaginaceae

Buck's-horn plantain seedheads

211

The Plantaginaceae family includes a wide range of different genera and species. One of the largest tribes is Veroniceae, though included here is just one member of that, the birds-eye speedwell. The forager's main focus with this family is one of the smaller groups, Plantagineae, more specifically the *Plantago* species, otherwise known as plantains.

The humble plantain can easily be overlooked because it is such a ubiquitous weed. However, it not only offers different types of nutritious food, but among the 20 or so species that have spread across the country, there are about nine native ones. As some of the natives are rare, I'm sticking with the most common and most useful species here.

Both plantain leaves and seeds are edible and high in calcium, vitamins C and K and beta-carotene, as well as various other nutrients. The seeds are a source of fibre and are used in psyllium. Although you are likely to find this weed all over the place, do not be tempted to forage close to roads, where pesticides might have been used or animals may have wandered — and rinse well.

Ivy-leaved toadflax

SCIENTIFIC NAME

Cymbalaria muralis

DESCRIPTION

A fast-growing, sprawling groundcover with small leaves similar to ivy but more rounded, that look like tiny indented lily pads. They are green on top and pinky-purple underneath. The flowers are similar to a snapdragon's in shape but tiny and purple with a splash of yellow and white in the centre and sometimes dark-purple lines on the upper two petals. They produce small circular fruit, which begins green, turning purply-brown then black.

DISTRIBUTION

Dotted through much of the country, particularly Auckland, Wellington and Christchurch, but not the southern half of the west coast of the South Island.

HABITAT

Happily grows out of shaded rock walls and pavement crevices, as well as invading a range of well-drained soils, including pot plants.

SEASONALITY

Flowers through summer, the leaves available for much of the year but dying back in winter. Best foraged in spring as the leaves become more bitter later in the year.

SIMILAR SPECIES

There may be confusion with other plants called toadflax — the *Linaria* species, which are related to *Cymbalaria muralis*.

Key edible uses

The leaves are high in vitamin C, mildly bitter and cress-like in taste, with a refreshing, slightly crunchy texture — they are well loved by hens. They make a good addition to salads and can also be blanched quickly and served as a vegetable. When black, the miniscule fruit can be eaten, but they are toxic when still green. The flowers can also be eaten as a garnish or frozen within an ice cube.

Opposite Ivy-leaved toadflax often grows in creeping mats over vertical surfaces.

Buck's-horn plantain

SCIENTIFIC NAME

Plantago coronopus

DESCRIPTION

Forming a rosette, its thick, narrow leaves grow up to 20cm long and have short, pointed teeth growing either side up the central rib of each leaf. The cylindrical spikes of tiny white flowers grow on upright stalks up to 45cm high issuing from the plant's central base. The seedhead of pink-brown seeds is not as thick as that of the broad-leafed or greater plantain (see page 216).

DISTRIBUTION

Widespread throughout coastal Aotearoa New Zealand.

HABITAT

Occurs along the edge of shorelines, often in areas with very well-drained soils and often close to rocks.

SEASONALITY

Grows throughout the year.

FORAGING TIPS

Best to gather the fresh leaves in spring after a period of rain. Can become very bitter in flavour after prolonged periods of dry weather.

Above opposite Buck's-horn plantain has thick and angular leaves. **Below** Plantain seed heads.

Key edible uses

The young leaves can be eaten raw or cooked. They are slightly bitter with a mild smoked-coconut flavour. They have a crunchy texture when raw, though if you prefer them to be more tender, briefly blanch them in boiling water then immerse in cold. The leaves are an excellent addition to salads or, especially if older, can be used as a side vegetable or added to soups or stews.

Greater plantain

SCIENTIFIC NAME

Plantago major

OTHER NAMES

Broadleaf plantain, cart track plant, common plantain, waybread, white man's footprint

DESCRIPTION

Growing up to 25cm long from a rosette, the leaves are large with a short stalk holding an oval-shaped green blade with distinctive vertical veins. The long spike holds tiny cream-green flowers that produce many purple-brown seeds.

DISTRIBUTION

Found throughout Aotearoa New Zealand but absent from higher-altitude areas above 1000m.

HABITAT

Often found in well-drained areas along edge environments in open spaces. A very widespread colonising species.

SEASONALITY

Best to harvest in winter and spring after rainfall when the leaves will be most tender, and in summer and early autumn for the seeds.

Clockwise from left Ripe seedheads; unripe seeds and leaf; mature plants; fresh leaf.

Key edible uses

The broad leaves have a similar bitter flavour to narrow-leaf plantain (see page 218) but the slightly bitter seeds are slightly larger. The seeds are rich in vitamin B1; they are suited for use in baking and especially tasty when added to a flatbread, but can also be lightly toasted and sprinkled into cereals or over salads. To avoid stringiness, remove the leaf stalk and stick to new-growth leaves, which can be wilted like spinach or blanched and added to salads. The young shoots are good raw in a salad or stir-fried in a little butter or olive oil to yield an asparagus-like flavour. The leaves can also be dried to make tea.

Narrow-leaf plantain

SCIENTIFIC NAME
Plantago lanceolata

OTHER NAMES
Buckhorn, English plantain, lamb's tongue, ribleaf, ribwort

DESCRIPTION
Growing in clusters from a central base, the thin, narrow leaves have distinct ribbed vertical lines running the length of the leaf, which can be up to 20cm long. The flower spike is shorter than in other plantains, more oval than cylindrical, and the tiny cream flowers extend on little stalks that circle just below the tip of the spike.

DISTRIBUTION
Widespread throughout Aotearoa New Zealand.

HABITAT
Often found in well-drained areas along edge environments, open grasslands and parks.

SEASONALITY
Available year-round but fresh growth of leaves in spring provides the best flavour.

Above opposite Narrow elongated leaves. **Below** Flowers forming into seed heads.

Key edible uses

The leaves have a mild, slightly bitter, coconut-like flavour and when young can be wilted lightly in salted water and served as a green. When dried they also make tea. The flower buds can be eaten fresh in salads. The seeds are a tasty and readily available superfood, and are great in salads but can also be lightly roasted and put into muesli, bread or other baking. When fresh they have a mild mushroom-like flavour.

Bird's-eye speedwell

SCIENTIFIC NAME

Veronica persica

OTHER NAMES

Common field-speedwell, Persian speedwell, winter speedwell

DESCRIPTION

A small, sprawling annual that sends out runners, it has deeply toothed oval leaves and blue flowers with a white eye and definite stripes on its four petals. The lower petal is often slightly smaller and paler than the rest, though this is not obvious as the flower is less than 1cm across. The flowers are borne singly on their stems but together form a mat dotted with blue. It produces heart-shaped seedpods.

DISTRIBUTION

Spread across all but the west coast of the South Island, and particularly prevalent in Auckland, Hamilton and Christchurch.

HABITAT

Found growing amidst moist grassy areas and wasteland.

SEASONALITY

Appears in spring with flowers for much of the year.

Opoosite The tiny but vividly coloured flower highlights this small plant.

Key edible uses

Harvest the new growth when it is less bitter. Leaves, stems and flowers can all be eaten: all added to tea; the leaves and shoots added to salad, smoothies, pesto and cooked vegetables; the flowers used as a garnish. Traditionally used medicinally, the plant is rich in vitamins and minerals.

Polygonaceae

Fresh growth sheep's sorrel

With a name that means 'many knees', probably because of the nodes or joints up the stems of many of its members, this group is also known as the knotweed, smartweed or buckwheat family. There are a couple of knotweeds included here, but the main interest for the forager is the buckwheat element. Like many *Amaranthus* species, this family includes pseudocereals that produce gluten-free seeds that can be used in place of grass-derived grains. Commonly called buckwheats, many of these plants can be foraged for their edible leaves as well as seeds.

The common buckwheat, *Fagopyrum esculentum*, is grown as a cover crop, but the buckwheats that can easily be foraged in Aotearoa New Zealand are the *Rumex* species, the sorrels and docks. Note that *Rumex* sorrels are distinct from wood sorrels (*Oxalis* spp.). *Rumex* leaves can also be eaten; these generally have a strong lemon flavour and are high in vitamin C as well as other beneficial nutrients. These contain oxalic acid (see page 12), but like most other plants that contain this toxin, you would need to eat an impossible amount to suffer adverse reactions.

Water pepper

SCIENTIFIC NAME

Persicaria hydropiper

OTHER NAMES

Polygonum hydropiper, Marsh pepper knotweed

DESCRIPTION

Grows 30–60cm high with a smooth stem bearing dark-green alternate, lance-shaped leaves up to 10cm long with almost invisible marginal hairs. The new shoots are red, turning green with red markings under the leaf nodes, where a papery sheath forms. Abundant narrow green-white-pinkish flower spikes appear at the ends of the stems.

DISTRIBUTION

A patchy distribution throughout Aotearoa New Zealand. Often in low- to mid-altitude locations, generally more common in northern areas and appears to be absent from Southland.

HABITAT

Found in areas with waterlogged soils, often bordering wetlands and streams. Mostly found in open environments but can tolerate partial shade.

SEASONALITY

Fresh growth appears in mid-spring, flowering from mid-summer to early autumn, depending on the location. When in flower the plant becomes more conspicuous. Dies back in winter.

CAUTIONARY NOTES/CONFUSION SPECIES

Although the plant has been used in traditional Chinese medicine, it is important to eat this species in moderation only, as large amounts over a prolonged period could result in adverse effects. The oils in the plant can be a skin irritant for some people.

It is similar to slender knotweed, *Persicaria decipiens*, and willow weed, *Persicaria maculosa*, but both usually have a single dark blotch in the centre of their leaves and their pinker flowers are borne in tight clusters at the top of the stems rather than spread down much of the top part of the stems. Neither is edible unlike the other lookalike Vietnamese coriander (*Persicaria odorata*), but is not as weedy.

Key edible uses

The leaves and stems have an aromatic coriander flavour followed by a pungent and peppery taste, leaving a warm, spicy sensation, similar to black pepper. Used fresh as a culinary herb and for adding to sauces and salads and, given its use in Japanese cuisine, also to Asian dishes. You can dry the leaves and powder them into a spice. In addition, the seeds can be blitzed to make a hot wasabi-like paste.

Opposite Water pepper lends its flavours well to a host of Asian cuisines.

Japanese knotweed

SCIENTIFIC NAME
Reynoutria japonica

OTHER NAMES
Fallopia japonica, *Polygonum cuspidatum*, Asiatic knotweed

DESCRIPTION
This has hollow, jointed, bamboo-like stems that can form thickets and grow over 2.5m high. Fresh-growth stems are red, becoming green as the plant matures. The leaves are broadly heart-shaped and arranged alternately along the stem. It has clustered small cream flowers on multiple spires and forms extensive root/rhizome underground networks.

DISTRIBUTION
Found scattered throughout Aotearoa New Zealand with strongholds on the northern part of the west coast of the South Island and the upper North Island. Has recently been eradicated from Stewart Island and its range may contract as further eradication programmes take place.

HABITAT
Likes damp edge-zone environments and often spreads along river courses. Can adapt to a wide range of soil types.

SEASONALITY
Available year-round but has fresh growth in late winter and early spring and flowers through the summer.

CAUTIONARY NOTES
Be careful not to spread seeds or cuttings as it is considered invasive.

Key edible uses

High in vitamin C and resveratrol, which has antioxidant properties, the fresh young shoots and leaves can be eaten as a vegetable, juiced or used as a rhubarb substitute, given that they have a similarly acid flavour. Older stems and leaves become fibrous, so peeling and cooking or else cooking and sieving is required. Seeds are also edible when raw or cooked, and the roots are also occasionally consumed. It is used in Japanese cuisine and in Japanese and Chinese medicine. Japanese knotweed is quite high in oxalic acid, (see page 12), so consume in the same amounts as you would rhubarb stems.

Opposite Japanese knotweed. **Inset** Distinctive mottled red stem.

Sheep's sorrel

SCIENTIFIC NAME

Rumex acetosella

OTHER NAMES

Field sorrel, red sorrel, sheep dock, sour weed

DESCRIPTION

This sorrel grows on the ground in clumps, with single leaves on a short, ridged stalk. The small leaves have a distinct pointed oval shape with two spur-like lobes at the base and grow up to only 8cm long. Red seeds form on a long central stalk in late spring and summer, which is when sheep's sorrel is most conspicuous.

DISTRIBUTION

Widespread in areas with well-drained soils throughout most of Aotearoa New Zealand, with strongholds in Otago and Canterbury. Rare north of Auckland.

HABITAT

Often found on the edge of riverbeds, or in areas of low grass or on track edges.

SEASONALITY

Best foraged in early spring for leaves and summer for seeds, but it is available throughout the year in small amounts and can have large flushes of fresh growth in autumn. Best to collect after periods of rain as leaves can become tough in dry conditions.

FORAGING TIPS

As it grows in clumps on the ground, thorough washing of the leaves is recommended, especially when collected from the edge of tracks.

Key edible uses

The citrus-flavoured leaves are best used when young and in moderation, added raw to smoothies, salads, pesto or salsas, or else cooked: sorrel soup is particularly good. The root can be dried and ground to be an addition to flour. The seeds are edible though too small to be an adequate cereal substitute, but they can be roasted and added to flour and other baking mixes. Contains oxalic acid, see page 12.

Above opposite Sheep's sorrel has a distinct red seed head. **Below** Sheep's sorrel leaves, best eaten when young.

Dock

SCIENTIFIC NAME

Rumex conglomeratus, Rumex crispus, Rumex obtusifolius

OTHER NAMES

Bitter dock, broad-leafed dock, butter dock, clustered dock, common dock, curly dock, dock leaf, round-leaved dock, sharp dock, yellow dock

DESCRIPTION

Most dock species form clumping rosettes with large veined leaves, which are lance-shaped with wavy edges for the clustered dock, slightly curled in curly dock and broad with round tips in the larger variety broad-leafed dock. As the leaves mature, they often get a rusty-red mottling and the stem can also become red. The flowers are small, green and inconspicuous, arranged in dense clusters on upright spikes, producing green seeds that are densely packed and turn a rusty-brown colour when mature.

DISTRIBUTION

Widespread, particularly at lower elevations.

HABITAT

Adapts to a wide range of soil conditions in open and particularly moist areas. Often found in open-edge environments bordering farmland. Can become invasive in farmland environments.

SEASONALITY

Available year-round but best harvested in winter and early spring when the green growth is at its most vibrant.

CAUTIONARY NOTES

Use the leaves in moderation because of the oxalic acid content.

Key edible uses

When small and young, the lemony leaves with their high vitamin and mineral content are well worth shredding raw into salads or steamed. Note that broad-leafed dock is blander than the others. Medium-sized leaves are best lightly blanched and served with vegetables or added to omelettes, stir-fries or soup. Larger leaves are bitter and tough but make an effective food wrap. The seeds can be harvested: when green add to salad; when brown treat as a salad grain or grind into a flour or coffee alternative or roasted whole add to breads, seed crackers and cereals.

Clockwise from left Common dock with seeds; curly (yellow) dock with seed heads; fresh growth clustered dock in flower; fresh picked dock leaves.

Climbing dock

SCIENTIFIC NAME

Rumex sagittatus

OTHER NAMES

Arrowleaf dock, potato vine, rambling dock, turkey rhubarb

DESCRIPTION

Grows as a creeping bush, with pointed leaves with sharp lobes that look like arrowheads. The leaves typically have prominent veins and can grow up to 15cm long. They are usually green but can sometimes have reddish or purplish tints, especially when heat-stressed. The flowers are pinky-white, followed by green seed clusters that become red.

DISTRIBUTION

In lowland North Island with strongholds around Wellington, Kāpiti Coast and Taranaki but also found in Hawke's Bay and Auckland. Has been recorded recently in Nelson and Marlborough regions, so it is likely to colonise the South Island where it is currently mostly absent.

HABITAT

Disturbed edge environments on bush edges, quarries and sand dunes. Most common on the edge of urban environments.

SEASONALITY

Available all year, but winter to spring is the best time to harvest the leaves. The seeds can be harvested over summer.

CAUTIONARY NOTES

Use the leaves in moderation because of the oxalic acid content.

Key edible uses

The leaves have a sharp sour flavour and make a good garnish with oily fish dishes. When young the leaves and stems can be cooked as a vegetable. The red seeds also have a good flavour and can be used as a grain in cereals and in baking.

Opposite Climbing dock leaves are pointed, and while usually green, can sometimes have red or purple tints.

Rosaceae

Cherry plum, *Prunus cerasifera*

This family is filled with multiple edible species. While seemingly diverse — ranging from small ground-huggers to shrubs to trees — many have serrated leaves and thorns and many, particularly the edible ones listed here, have a similar flower. If you think of the typical flower of the dog rose — the simple older variety before hybridisation turned roses into multi-petalled blooms — there are five pale petals around a centre of multiple yellow stamens. The buds open into a cup before flattening out. The flowers are often white and/or pink. This delicate flower is similar in other Rosaceae species, from the blossom of *Prunus* and strawberry to that of apple and pear, *Pyrus communis*.

Note that the seeds in the Rosaceae family — including apple pips — often contain small levels of hydrogen cyanide; it would be very difficult to consume a harmful amount, as very few people consume the seeds and pips in excess.

Many introduced plants have escaped from the gardens of early settlements and farms or might be found beside old roads, rivers or train tracks. Those from our orchards are easily recognisable and we need little advice on how to recognise and eat them. However, they may well develop different forms from those carefully grafted and hybridised varieties available in garden centres, the fruits perhaps smaller or tarter than those we buy in the supermarket. And while apples, *Malus domestica*, are easy to recognise, crab apples might also be found in a couple of places (such as *Malus baccata*, *Malus coronaria*, *Malus ioensis*, *Malus prunifolia* and *Malus sylvestris* in the Canterbury region and *Malus toringo* around Auckland), producing sour fruit that might look more like cherries, but is best cooked into jellies and jams.

Like apples, the *Prunus* species are another member of the Rosaceae family that are well known from our gardens, but may also be a target for foraging. Again, they produce fruit we are familiar with but also variations that are less common. They nearly all have white or pink blossoms with the typical five petals. The leaves are generally similar, oval — some more lance-like than others — with finely serrated margins.

Plums (*Prunus domestica*) come in a range of colours, some with purple flesh and skin, some with purple skin and yellow flesh, some with a reddish or yellow skin and yellow flesh. Then there are far less common damsons (*Prunus insititia*) and sloes or blackthorns (*Prunus spinosa*), which are similar to everyday plums but dark purple, slightly smaller, tart in flavour and best used for cooking with meat, or in jams and chutneys, or for making wines and liqueurs.

Cherry plums (*Prunus cerasifera*) are between plums and cherries in size and characteristics, and like sloes are better cooked with sugar. Wild cherries (*Prunus avium*) are spreading across the country and are well worth foraging, but you might also encounter sour cherries (*Prunus cerasus*). These are also edible and very similar to *Prunus avium* to look at — though the tree is smaller, the fruit darker, the stalks shorter — but as their common name suggests are best cooked with sugar. *Prunus serotina*, the black or rum cherry, is from America, and is also beginning to naturalise here and is also edible. In addition, a number of cherries — particularly Japanese varieties — fill our streets and parks having been cultivated for their spectacular blossoms rather than their fruit, which is generally small, very bitter and not worth foraging. The blossoms, however, often have an aromatic or marzipan-like flavour, which can be brought out through infusing in spirits or syrup to make alcoholic drinks or sorbets.

Then there are apricots, *Prunus armeniaca*, which you are less likely to find in the wild though they are easily recognisable if you do.

Clockwise from left Mirabelle plums; damson plums; luisa plums; plum blossoms.

Chinese quince

SCIENTIFIC NAME

Chaenomeles speciosa

OTHER NAME

Flowering quince

DESCRIPTION

A thorny deciduous shrub, growing up to 2m high, with small oval leaves. The clustered flowers have five rounded petals around yellow stamens and occur in a range of colours, often red or salmon-pink but also white. After flowering the shrub produces small, rounded, yellow fruit.

DISTRIBUTION

Found throughout Aotearoa New Zealand, often in areas near habitation but occasionally in the wild in abandoned edge zones, farmland and river margins.

HABITAT

Often in areas with damp fertile soils. Tends to like open environments but also tolerates shaded locations along edge zones.

SEASONALITY

Flowers from mid-July to September with ripe fruit over summer.

FORAGING TIPS

Harvest after a cold snap as this softens the fruit.

SIMILAR SPECIES

A similar, smaller variety is the Japanese quince, *Chaenomeles japonica*. It is mostly grown within gardens, but is worth foraging if you do find it has wandered outside into wasteland. It is a thorny shrub, only a couple of metres high, with the same simple flower except that it is vivid orange, coming out both in winter and summer. The leaves are oval with slightly serrated margins. Its fruit is similar to European quince, *Cydonia oblonga*, but is a lot smaller, and can be eaten in the same way, but not raw, and like a quince it needs to be bletted (see page 242).

Key edible uses

The flowers are edible and make an attractive garnish or can be infused in oils or vinegars or candied. The fruit is very high in vitamin C but hard and very acid when raw, so is best stewed with sugar and sweeter fruit to make jam, jelly and preserves.

Opposite Chinese quince has distinct, bright-red flowers that can be candied. **Inset** The fruit is smaller than common quince, and should be bletted.

Hawthorn

SCIENTIFIC NAME
Crataegus monogyna

OTHER NAMES
Haw, may, hedgethorn, whitethorn

DESCRIPTION
This deciduous tree has spindly, spiked branches, rough bark and profuse white or pink blossom that produces rounded oblongs of dull-red berries. The leaves are dull green, each indented into five rough lobes (two bigger lobes, two smaller above and one at the top), which in turn are toothed. The tree can grow up to 10m high. There is a slightly smaller species, *Crataegus laevigata*, with leaves that have less-defined lobes and seedier berries.

DISTRIBUTION
Widespread, but more common on the eastern side of the South Island.

HABITAT
Often in hedgerows along roadside margins in drier parts of the country. Also grows wild on open hillsides in areas of rough farmland.

SEASONALITY
All year in flower, but with peak flowering in spring and berries in summer and autumn. Berries often stay on the tree through winter too.

CAUTIONARY NOTES
Watch out for the thorns and be cautious if you are on medication.

SIMILAR SPECIES
The Mexican hawthorn, *Crataegus gracilior*, has similar flowers but oval serrated leaves. It produces yellow haws that are larger than the European species, but as lovers of warmer climates it has not spread. The Indian hawthorn, *Rhaphiolepis indica*, is a completely different species, yet still a member of the Rosaceae family; it has purple-black berries, which are edible but only when cooked. Often used as a hedge or landscaping plant, it has spread across much of the country, particularly in the North Island. The whitebeam, *Aria edulis* or *Sorbus aria*, produces white blossom and red berries similar to *Crataegus monogyna*, though the leaves are not lobed like the hawthorn. Its berries are also edible, but only when almost rotting (see bletting, page 242).

Opposite Hawthorn berries ripening. Full ripe berries are a deep red colour. **Inset** Hawthorn in bloom.

Key edible uses

Full of antioxidants, vitamin C and beneficial minerals, the plant is used for medicine as well as a source of food, having edible flowers, leaves, shoots and berries, which can be eaten raw or cooked, though cooking improves the flavour. The nutty new shoots can be chopped into salads along with a garnish of the blossoms. Dried leaves make a good tea, while the seeds can be roasted as a coffee substitute. The aromatic flowers can be added to syrups and desserts, but it's usually the berries that are used for a variety of recipes both sweet and savoury. The berries are best strained to remove the seeds, and they can be dried whole and ground to be mixed with flour for baking.

Medlar

SCIENTIFIC NAME
Crataegus germanica

OTHER NAME
Mespilus germanica

DESCRIPTION
A shrubby deciduous tree that can grow up to 8m high but is generally smaller. The leaves are oval with hairy undersides, looking a bit like apple tree leaves, and have tiny, almost imperceptible serrations. With heart-shaped petals like the dog rose, the white or pink flowers are similar in form to many other members of the Rosaceae family and are quite large, up to 5cm, and produced singly, much more spaced out than on a hawthorn. The fruit looks a bit like largish fuzzy brown rosehips with a similar opening where the sepals were.

DISTRIBUTION
A rare find, dotted around the country but only found in urban areas.

HABITAT
Barely escaped over the garden fence.

SEASONALITY
Flowering in late spring and early summer, the fruit appears in autumn and winter.

Key edible uses

Although not common, it is worth including this fruit here partly for curiosity value and partly to explain how some foraging items need bletting. Some fruit is hard and not edible when picked, even when ripe, but when left on the tree over winter or else picked but stored for a period of time, to the verge of rotting, the acids and tannins break down and a sweeter flavour develops — making it not only edible but often delicious (see also quince and rowan, though persimmon is another example that can be eaten when hard but is much sweeter when left to soften). Once bletted, the medlar will look unappetising, but think of it as apple sauce, which can then be eaten raw or cooked into crumble or other desserts, or made into jelly or used as a lemon substitute in lemon curd. The fruit is high in potassium and other minerals and was used in traditional medicine.

Quince

SCIENTIFIC NAME
Cydonia oblonga

DESCRIPTION
Growing up to 8m tall, this is another deciduous tree. The leaves are broader than the apple, spade-shaped with smoother margins, which are slightly hairy. The familiar five-petalled blossoms come in white or pink, looking quite similar to the medlar (see page 242) in being large and produced singly. The fruit begins as fuzzy globes, becoming more like the skin of a pear, growing larger than both an apple or a pear and looking like a bulky, misshapen combination of both. It turns from green to yellow as it ripens. When cut across the middle widthways, the central core forms a five-pointed star, which may be where the quince (meaning five) got its name, though that is a matter of debate.

DISTRIBUTION
Adapted to a range of climates, mainly from Christchurch northwards.

HABITAT
Found escaping not far beyond urban areas.

SEASONALITY
Flowers in spring, fruits over autumn.

Opposite Quince fruit are bright yellow and have a slightly furry texture. It has many culinary uses, both sweet and savoury.

Key edible uses

Containing vitamin C, potassium and other minerals, this is another fruit that needs some amount of bletting to become edible (see page 242), otherwise it is too hard and astringent to be palatable. Leave until the skin has become yellow and the flesh slightly softer. Although there are some varieties that can be eaten raw, especially when grown in warmer areas, the fruit is usually best cooked, particularly as the raw texture is often grainy. The aromatic, tart fruit can be cooked with meat in savoury Mediterranean dishes or else combined with honey or sugar and baked, poached, stewed, added to pies (particularly apple or pear) or made into a sauce or pickle. Given that they are high in pectin, quinces can also be turned into paste, marmalade, jam and jelly. In some European countries quince is used to make alcoholic drinks, a simpler version of which is easy to do by steeping grated quince with a bit of sugar in gin or vodka and leaving for a few months in the back of your pantry, shaking every so often. The flowers are also edible and can be used as a garnish.

Loquat

SCIENTIFIC NAME
Eriobotrya japonica

OTHER NAMES
Rhaphiolepsis bibas, Japanese loquat

DESCRIPTION
This member of the Rosaceae family is an evergreen tree growing up to 8m tall, though is often smaller. It has hands of long thick, dark-green, oval leaves with pointed ends, well-defined parallel veins and subtly serrated edges. The aromatic white flowers have the usual five petals, these being well spaced and borne in clusters along short stems. The fruit are yellow or orange or sometimes reddish and are downy, looking a bit like small apricots. Each contains a large shiny brown seed, though some varieties have more.

DISTRIBUTION
Spread by birds, this is a pest plant in Auckland northwards. It can also be found across the North Island and in a few mostly northern areas in the South Island.

HABITAT
Spreads easily into bush and wasteland.

SEASONALITY
Flowers April to September, fruits October to December.

Opposite Loquat fruit is delicious straight off the tree, but also makes a great jam due to its high pectin levels.

Key edible uses

High in vitamins and minerals, as well as sugar, the fruit is sweet, slightly tart and juicy fresh off the tree, and once deseeded can be added raw to fruit salad and smoothies or made into a juice, or even fermented into wine or infused in spirits. Being high in pectin, it will make good jam, marmalade and jelly, but can also be poached or cooked in tarts or made into chutney or sauces. A coffee substitute is made from the roasted seed, — this is another plant that contains cyanogenic glucosides (see page 203) but any hydrolysed cyanide would be destroyed by toasting the seeds.

Woodland strawberry

SCIENTIFIC NAME

Fragaria vesca

OTHER NAMES

Alpine strawberry, European strawberry, wild strawberry

DESCRIPTION

We all know what a garden strawberry is like, but those modern varieties are hybrids developed from earlier wild forms, one of which is the woodland strawberry. Borne in threes, the toothed leaves have well-defined parallel veins. The five-petalled flowers are white with yellow stamens and a yellow mounded centre. The red fruit is small, covered in yellow-red seeds on the outside.

DISTRIBUTION

Found throughout Aotearoa New Zealand but absent from higher-altitude areas, above 1000m.

HABITAT

Sends runners through damp shady margin areas, often found under woodlands or in foothill areas. Tends to be absent from coniferous forests.

SEASONALITY

Flowers in spring and fruits in summer through to April.

FORAGING TIPS

Take a small container to pick berries into to prevent them from getting crushed.

CAUTIONARY NOTES

Anyone allergic to shop strawberries will also be allergic to these.

SIMILAR SPECIES

The mock strawberry (*Potentilla indica*) has similar leaves and flowers, though the latter are yellow and the central stamens rise from a yellow mound. Its small strawberry-looking fruit is edible but often quite dry and has a very mild flavour, so while it can be eaten in jams or as a garnish it is not as desirable as the woodland strawberry.

Key edible uses

The sweet aromatic fruit is best eaten raw, but can be dried or added to desserts and preserves, though finding enough to make a substantial dish on its own might be a challenge. The leaves can be used fresh or dried to make herbal tea.

Above opposite The small woodland strawberry has a concentrated flavour. **Below** The foliage grows in mats.

Rosa species

Roses need little introduction except that their edible qualities are often overlooked. In the wild, you'll often find them in thickets, sprawling under the sun over roadside banks or in edge environments. They mostly flower in spring and summer, producing hips in summer and autumn, which are especially distinct over the winter months when the leaves have dropped.

The species that has spread most readily into the wild in Aotearoa New Zealand, particularly in the South Island, is the sweet briar, *Rosa rubiginosa*. This deciduous shrub can grow up to 2m high, with its arching branches covered in small, in-curving thorns. The leaves consist of several smaller leaflets with serrated edges. The flowers are typically pale pink with the familiar five petals and have a very sweet fragrance. The often-prolific fruit known as rosehips begin green, turning to scarlet.

Another old rose that has naturalised here is the dog rose, *Rosa canina*, which can grow even higher than sweet briar and produces larger hips that are both sweet and tart. But you may also come across other scramblers, like *Rosa multiflora* or the sometimes bright-pink *Rosa lucieae*, among a number of different varieties that have jumped the fence. The hips are typically red or orange and have a sharp tart flavour.

Use gardening gloves when you forage for roses, as the thorns can be savage. And make sure you make the most of this versatile plant as the flowers, leaves, fruit and seeds can all be consumed.

The petals have a sweet floral flavour and can be used as a garnish, candied with sugar as a decoration, used to make a fragrant jelly or jam, or dried and made into a herbal tea. The base of the petal might be bitter, in which case slice it off. The mildly flavoured dried leaves can also be used for herbal tea. The ripe hips are high in vitamin C, but they have fine internal hairs that can cause irritation to the mouth or throat. This can be overcome by boiling the hips whole and using a fine sieve to separate the juice from the skin and seeds. The juice, which contains good levels of pectin, is commonly used to make cordial, syrup, jam, jelly and sauce. Alternatively, the hip can be dried and used in herbal tea or else ground into a powder and added to flour for baking; the seed is high in vitamin E. If you have the patience to carefully remove the outer flesh from the seeds and their protective hairs, you can turn the chopped pulp into fruit leather either in a dehydrator or by spreading it into a thin layer on baking paper and leaving it overnight in the oven on its lowest setting.

Briar rose flowers

Bush lawyer

SCIENTIFIC NAME
Rubus australis

MĀORI NAMES
Akatātarāmoa, tātaraheka, tātarāmoa, taraheke

OTHER NAME
Swamp lawyer

DESCRIPTION
A delicate climber with tiny hooked thorns and small, toothed, green leaves that have distinctive dark-purple veins across the surface. The tiny white flowers are scented, producing fruit that looks like yellowy-red blackberries dangling from the vine in clusters.

DISTRIBUTION
Found throughout much of the country, but less in eastern regions.

HABITAT
On lower-altitude bush areas, where it can climb up trunks.

SEASONALITY
Flowers early spring to summer; berries appear from late spring into autumn.

SIMILAR SPECIES
There are various other similar species of native bush lawyer: *Rubus cissoides*, *Rubus parvus* (creeping lawyer), *Rubus schmidelioides* (white-leaved lawyer) and *Rubus squarrosus* (leafless lawyer). These all have longer, narrower leaves but similar flowers, hooks and climbing habits and the yellow-red edible berry. The wine raspberry (*Rubus phoenicolasius*) is another Rubus species that produces edible berries, though these are more like small raspberries that begin yellow and turn bright-red then dark-red when ripe. Scrambling up to 4m, it has red bristles on its branches as well as erect thorns. The leaves are rounded with a point at the end and notched margins, appearing in threes, their underside covered in fine white hairs. They are not yet widespread but can be found close to settlements, mostly from Hamilton to Kaikōura. Sweetly tart, the juicy fruit has a raspberry flavour and can be eaten raw or cooked into pies, jams, and syrups.

FORAGING TIPS
Take a small container to pick berries into to prevent them from getting crushed.

Key edible uses

Mildly tart in flavour, the edible berries can be eaten raw or cooked, which makes them sweeter. They can be used in jam or desserts, though as they are not strongly flavoured are best combined with other fruit.

Above opposite Bush lawyer fruit forms in loose clusters, turning red, then orange as it ripens. **Below** Bush lawyer flowers are bright white.

Blackberry

SCIENTIFIC NAME
Rubus fruticosus

DESCRIPTION
This is another family member that needs little introduction. Grows as a creeping bush with five to seven lobes of serrated green leaves on each stalk, and dense spines on the branches. It has those familiar white, or sometimes pink, flowers, which turn into red then black berries that form in small clusters when fully ripe.

DISTRIBUTION
Throughout Aotearoa New Zealand in a wide range of edge environments.

HABITAT
Mainly in open and edge environments and most common around urban centres. Requires open sunlight but will grow in forest understories where there is moderate light penetration.

SEASONALITY
Berries can be harvested from late January to mid-April; leaves are best foraged in spring.

FORAGING TIPS
As this plant often grows as a dense thorny shrub, typically on sloping ground, good footwear, protective clothing and gloves are recommended. A hand-saw can be useful for making pathways through the brambles when harvesting. Also have a small container to pick berries into to prevent them from getting crushed.

CAUTIONARY NOTES
As blackberry is considered an invasive weed, be careful to avoid plants that have been sprayed with herbicides.

Key edible uses

The ripe black berries are primarily harvested to eat raw, juiced or cooked in pies, jams or syrups. A herbal tea can be made from the fresh leaves in the spring, though some people with sensitive stomachs can react badly to the tannins in the leaves. The new shoots can be peeled and chopped into salads.

Inset Blackberries in various stages of ripeness. **Clockwise from left** Ripe blackberries; white-purple blackberry flowers; blackberry leaves have five-seven serrated green leaves on each stalk. Wear gloves when foraging, as it usually grows as a thorny shrub.

Salad burnet

SCIENTIFIC NAME
Sanguisorba minor

DESCRIPTION
This small plant has dark-green compound leaves made up of small lobes with definite serrations growing up either side of the rib, giving them a delicate fern-like appearance. The stems are slender and grow upright to 30cm high. Contrary to so many of the Rosaceae family, salad burnet produces small, round, globe-like flowerheads that are composed of tiny, densely packed, red-to-maroon petals. The plant has a strong cucumber-like scent.

DISTRIBUTION
Found at scattered sites mainly around Christchurch and Wellington.

HABITAT
Primarily a garden escapee, found on the edges of gardens or abandoned edge zones near habitation. Often grown as a bedding plant.

SEASONALITY
Fresh growth is available in spring and autumn. It tends to die back over the winter months.

FORAGING TIPS
Fresh growth is best, as older and more woody plants have more of a bitter flavour.

Key edible uses

The leaves have a refreshing cucumber flavour and are often used fresh as a palate-cleanser, added to green smoothies, herb butter, salads or dressings or used as a garnish. The leaves can also add their aromatic flavour to a wide range of dishes, such as soups and sauces. Dried leaves also make herbal tea.

Rowan tree

SCIENTIFIC NAME
Sorbus aucuparia

OTHER NAME
Mountain ash

DESCRIPTION
A medium-sized tree growing 4–10m high with a rounded shape and widely spread branches. The leaves are dark green in colour with serrated leaflets growing in pairs up a central stem. Over the autumn the trees have a striking array of colours from yellow to orange and red. The tree has distinct clusters of white flowers that produce red berries that ripen over the summer and will stay on the tree into the early winter.

DISTRIBUTION
Widespread throughout the southern half of the eastern and central South Island. Rare in the North Island but occurs at some locations in the Central Plateau. Typically prefers cooler climates.

HABITAT
Often used in ornamental tree plantings in urban environments, it has spread to roadsides and river margins.

SEASONALITY
Rowan berries are available from late summer to the end of autumn.

FORAGING TIPS
Having a ladder can be useful for accessing higher fruit. Pick the berries after a frost or cold period as the lower temperature helps concentrate the flavour. The berries must be cooked to be made edible.

Key edible uses

The berries are sour and must be cooked or frozen before consumption, but they are high in vitamin C and enhance rich meat dishes. They can be used to make jam or jelly that goes well on a cheeseboard, or syrup.

Rubiaceae

Ripe South African boxthorn flowers

The best-known species in this family are *Coffea* spp., which produce coffee beans. Interestingly, other members of the family may look quite different but their seeds often give off a coffee aroma when roasted, and some even have a slight coffee taste when drunk as an alternative. With that exception, Rubiaceae isn't widely known as a source of food; though it has a number of members valued for their medicinal purposes, such as *Cinchona* spp., from which quinine is derived.

In this country, foragers focus on two very different offerings: berries from *Coprosma* and foliage from cleavers.

Coprosma

SCIENTIFIC NAME

Coprosma spp.

MĀORI NAMES

Hukihuki, hūpirau-ririki, hūpiro, kanono, kapukiore, karamū, karangū, kawariki, māmāngi, mingi, mingimingi, manono, patutikitiki, raurēkau, tātaraheke, taupata

DESCRIPTION

There are over 50 different native *Coprosma* species in Aotearoa New Zealand, ranging from shrubs to small trees, but mostly encountered as scrubby sprawling bushes with thin branches growing to a height of 1–3m. Their flowers are generally indistinct, and the leaves can vary from narrow spikes to tiny discs to broad ovals, which makes identification tricky. Depending on the species, they produce edible orange, red, pink, blueish-grey, purple or white berries, which are slightly translucent, sweet-flavoured and contain two seeds.

FORAGING TIPS

Although birds love *Coprosma* berries so spread the seeds widely, there are a few that are at risk, such as the sand coprosma *Coprosma acerosa*, so target the common ones, such as the species described in the following pages. Use fruits only when fully ripe and soft.

CAUTIONARY NOTES

There are other shrubs in Aotearoa that produce berries which are toxic — such as the purple berried tutu, *Coriaria arborea*, or red-berried *Cotoneaster* spp. — so correct identification is essential. Get to know the poisonous plants, given that coprosmas are so variable.

Key edible uses

Some species have tastier berries than others but all are reported as edible. The roasted seeds have a coffee aroma and can be used as a substitute, but you would need to collect a lot of berries to make this worthwhile.

Above opposite *Coprosma repens*
Below *Coprosma propinqua*

Mingimingi

SCIENTIFIC NAME
Coprosma propinqua

DESCRIPTION
Note that there are other, completely different shrubs also called mingimingi. The coprosma mingimingi is a scrubby, sprawling bush with multiple thin branches, a rounded form and tiny strap-like leaves growing to a height of 1–3m, producing small (2–4mm), sweet-flavoured, blueish-grey or blue-flecked berries.

DISTRIBUTION
Found throughout Aotearoa New Zealand from coastal to alpine regions, generally in association with the edges of native bush, lake and river edges and coastal margins, or else in ornamental plantings.

HABITAT
Adapts to a wide range of habitats — both swampy and dry — with exposure to high levels of sunlight.

SEASONALITY
Berries are most prolific in late summer and autumn.

FORAGING TIPS
Look carefully on lower branches where birds have overlooked the ripe berries.

Inset Mingimingi has a sweet blue-grey fruit that make an excellent snack while foraging!

Key edible uses

The ripe berries have a sweet flavour and medium-sized seeds. They make a tasty snack, especially when tramping, and given their colour make an interesting garnish.

Taupata

SCIENTIFIC NAME

Coprosma repens

OTHER NAMES

Looking-glass plant, mirror bush

DESCRIPTION

A shrub typically less than 5m high, with distinctive rounded, very glossy, dark-green, leathery leaves that grow in pairs up the branches. The creamy-yellow flowers are unobtrusive and produce bunches of pale-orange berries of 4–6mm growing tightly off the branches. Very similar to karamū (page 264) but has a much larger berry.

DISTRIBUTION

Throughout coastal Aotearoa New Zealand and lowland areas, generally less than 100m above sea level.

HABITAT

Tolerates a wide range of coastal and forest edge environments, preferably with well-drained soils and lots of sunlight. Often also grown as an ornamental tree.

SEASONALITY

Produces orange fruit from February until May.

FORAGING TIPS

The juice can be collected from the berries by squeezing through a sieve. Use clippers to remove bunches of berries.

Key edible uses

The berries produce a sweet orange-coloured juice, which is useful for flavouring oils or vinegar or reduced to make a tasty syrup that can be drizzled over desserts. The seeds can be roasted to make an indigenous coffee and the leaves will make herbal tea.

Karamū

SCIENTIFIC NAME

Coprosma robusta

DESCRIPTION

A shrub typically less than 5m high with thick, glossy leaves that are pointed ovals, pale-yellow filament flowers and orange-to-red berries growing off the branches. The berries are similar to those of taupata (page 263) but smaller, generally darker and more oval in shape.

DISTRIBUTION

Found throughout Aotearoa New Zealand except in high alpine zones.

HABITAT

Because the seeds are commonly eaten by birds, karamū is very widespread in lowland forests and scrub throughout the country and occasionally some distance inland, but overall it is considered to be a coastal shrub.

SEASONALITY

Ripe berries have a long season, from January to May.

Key edible uses

The berries produce a sweet orange-coloured juice, which is useful for flavouring oils or vinegar or reduced to make a tasty syrup that can be drizzled over desserts. The seeds can be roasted to make an indigenous coffee and the leaves will make herbal tea.

Inset and opposite *Coprosma robusta* fruit and bud galls of the *Coprosma* specialist native mite *Acalitus cottieri*.

Cleavers

SCIENTIFIC NAME
Galium aparine

OTHER NAMES
Catchweed, goosefoot weed, sticky willy

DESCRIPTION
Often growing in dense beds, this distinctive plant has a long central stem and elongated, rounded leaves growing out from the stem in clusters like sunbursts, all covered in fine hooked hairs. It has a Velcro-like texture (hence some of the nicknames), tiny green-white star flowers and rounded, sticky seed burrs.

DISTRIBUTION
Found throughout the country, most commonly around urban centres and agricultural areas.

HABITAT
Common in edge zones in sheltered moist environments where there has been some soil disturbance.

SEASONALITY
Winter and spring, dying away in the warmer summer months.

FORAGING TIPS
Can easily be collected in bunches with a sharp knife.

CAUTIONARY NOTES
Some people get skin irritation from contact.

Wash thoroughly. Often it will have other green plants growing in among its beds, so take care to remove any 'imposter' plants.

Key edible uses

Although used in traditional medicine and high in vitamin C and various minerals, its sticky texture and sometimes bitter flavour makes it less appealing for eating raw; however, the young winter shoots have the best flavour. To avoid the hairiness, add to a green smoothie or pesto mix, or cook as a side vegetable or add to soups and stews. The steeped leaves give off a pleasant cucumber flavour, making a refreshing tea. The seeds can be roasted and ground to make coffee.

Opposite Cleavers often grow in mats, with long stems, but can also grow creeping along the ground.

Solanaceae

Poroporo plants with ripening fruit

Also called the nightshade family, Solanaceae includes such familiar edible fare as potatoes, tomatoes, eggplants, tamarillos, capsicums and chillies. The family is known for containing alkaloids — some of which are desirable, such as capsaicin which gives chillies their spicy heat; others are medicinal; while others are toxic, such as nicotine, which although can be medicinal, is highly toxic if ingested; and solanine. Solanine is present in both potatoes and tomatoes, but in such small doses that it has no effect on humans — unless you eat green potatoes. Thus, the forager has to walk a tightrope between useful edibles and poisonous undesirables when searching for members of this family.

Some plants will teeter on either side. For instance, I have included boxthorn and have never had a bad reaction to eating its berries, which are generally deemed edible; but the levels of alkaloids in its leaves could also be high in the berries and make it too much of a risk to forage. Likewise, poroporo berries are toxic when green but fine when ripe: it is therefore essential to make informed decisions, to taste in small quantities first and not to consume in pregnancy or if unwell.

Meanwhile, there are other Solanaceae members that we know to be poisonous and must not be consumed. These include angel's trumpet, *Brugmansia* x *candida*, thornapple, *Datura stramonium*, apple of Peru, *Nicandra physalodes*, and the berries of woolly nightshade, *Solanum mauritianum*. There are also a number of cherry-tomato lookalikes: the berries of false Jerusalem cherry, *Solanum diflorum*, bittersweet, *Solanum dulcamara*, apple of Sodom, *Solanum linnaeanum*, and Jerusalem cherry, *Solanum pseudocapsicum*. So proceed with caution; after all, this is the family of deadly nightshade.

While not all members of this family have similar flowers, there are some features that crop up in many species, particularly those we want to target for foraging. The tomato-type flower usually has five pointed, star-like petals that stretch backwards as the flower matures to expose its prominent central stamens. The flowers with this form are repeated in, for example, the nightshade species, and are generally white, yellow or purple or a combination, the stamens often contrasting with the petal colour. The potato-type flower is similar but the petals are broader, becoming broader still in a few other species, such as poroporo, but again the petals will lean back to let the stamens protrude.

Boxthorn

SCIENTIFIC NAME
Lycium ferocissimum

OTHER NAME
African boxthorn

DESCRIPTION
Growing up to 4m high as a dense bush and often in thickets, it has tangled thorny branches and small green leaves. Its small green trumpet flower opens to five purple petals around its stamens. The inner parts of the petals are each marked with a darker purple triangle, and as the flower matures the petals become paler and bend back to expose the stamens. It produces orange berries that are about 8mm across.

DISTRIBUTION
Throughout coastal Aotearoa New Zealand but absent from the west coast of the South Island.

HABITAT
Open environments with well-drained soils.

SEASONALITY
Fruit available from January to April.

FORAGING TIPS
It is best to use gardening gloves when picking the ripe fruit to avoid being pricked by the thorns.

CAUTIONARY NOTES/CONFUSION SPECIES
The leaves are toxic, so don't eat them. The berries are edible, but only eat a handful at a time as some people may get stomach upset.

Do not confuse with other red-berried shrubs, such as *Cotoneaster* spp., though the latter does not have thorns.

Key edible uses

Rich in vitamins and minerals, the ripe orange berries can be eaten raw and have a tangy tomato flavour.

Opposite Ripe boxthorn berries are bright orange and can be eaten raw.

Cape gooseberry

SCIENTIFIC NAME

Physalis peruviana

DESCRIPTION

Grows as a soft-stemmed bush up to 80cm high with slightly furry green leaves with a rounded base and pointed tip. The leaves develop a distinct dark veining as they mature. The dull-yellow flower has a brown blotch on the inner base of each of the five large petals, which open out into a trumpet and bend backwards as they age. The round fruit grows inside a soft outer case that turns papery and a buff colour when the fruit inside has become orange and mature. The fruit grows to about 1.5–2.5cm across.

DISTRIBUTION

Mainly appears in coastal regions from Christchurch northwards, generally near areas of settlement. Increases in abundance in northern regions and likely to increase in distribution and abundance with global warming.

HABITAT

Generally close to garden environments and sheltered, frost-free sites with well-drained but fertile soils. Also found on edges of native bush patches.

SEASONALITY

Peak fruit harvests in autumn, though in northern areas can produce fruit year-round. It flowers in spring, making the otherwise nondescript plant conspicuous.

FORAGING TIPS

Plants can conceal themselves quite well with the fruit often hidden under the large broad leaves, so look carefully along rank vegetation edges where they often grow.

CAUTIONARY NOTES

Only eat the fully ripe fruit and do not consume the outer case or any other part of the plant, as all parts except the fruit are toxic. The seeds are safe to eat.

Key edible uses

The bright-orange fruit has a sweet, slightly tangy flavour and contains lots of small edible seeds (a bit like a tomato). It is high in vitamins, phosphorus and other minerals. It is delicious eaten raw or cooked in jams, chutneys, pies and other desserts. The fruit can also be dried.

Above opposite Cape gooseberry flowers have brown inner marking. distinct markings. **Below** These cape gooseberries are not quite ripe as you can tell by the colour of the still green outer case. **Inset** Ripe fruit.

Poroporo

SCIENTIFIC NAME
Solanum laciniatum

OTHER NAMES
Kahoho, pōporo, tūpurupuru

DESCRIPTION
A soft-wooded shrub typically growing to 2–3m high, with leaves that are sometimes long spears, other times deeply lobed with angular points. Its purple, ruffled flowers have bright-yellow stamens. They produce large oval berries about 15mm long, which begin green, turning to yellow and then orange as they ripen.

DISTRIBUTION
Widespread throughout coastal and lowland Aotearoa New Zealand but rare on the west coast of the South Island.

HABITAT
Often colonises disturbed environments around areas where forest has been cleared or on bush edges where it can get sunlight.

SEASONALITY
Ripe fruit is available from September to May with a peak from February to April.

FORAGING TIPS
Ripe fruit is often best gathered close to the ground as the higher fruit tends to get eaten quickly by birds.

CAUTIONARY NOTES/CONFUSION SPECIES
The plant, flowers and unripe fruit, when green or yellow, are all toxic. Only consume when the fruit has turned orange.

Solanum aviculare is very similar but the flowers are often white or pale-purple; as these plants are at risk, they should not be foraged.

Key edible uses

The fruit is only edible when bright orange, soft and starting to split. It can be eaten raw or made into jam or chutney.

Inset Unripe poroporo fruit. Make sure you eat the fruit only when it bright orange, and ripe. **Opposite** Poroporo has spear-like leaves, with delicate purple flowers.

Black nightshade

SCIENTIFIC NAME
Solanum nigrum

DESCRIPTION
Black nightshade is a herbaceous plant that can grow up to 1m high. The leaves are pointed and vary: shaped either like arrowheads or ovals, sometimes irregularly toothed, with a matt-green colour that tinges purple under strong sunlight. The five petals on the small star-shaped white flowers bend back to reveal prominent yellow stamens in the centre. They are typically arranged in small clusters and produce small round berries that start green and ripen to shiny black. These berries are about 4–7mm across and contain numerous small seeds.

DISTRIBUTION
Widespread throughout Aotearoa New Zealand except in heavily forested and alpine environments.

HABITAT
Uses a wide range of habitats from partial shade to open sunlight. Also tolerates an extremely wide range of soil types.

SEASONALITY
Can flower and fruit throughout the year, but most berries tend to be ripe from late summer to late autumn.

CAUTIONARY NOTES/CONFUSION SPECIES
There is much debate about the safety of this plant, but it is widely consumed and generally accepted that only the unripe fruit and older leaves have dangerous levels of toxins. You should only consume the berries when they are fully ripe and purple-black.

Take care not to confuse with other nightshades, many of which are poisonous. These, though, are mostly distinct, for example hairy nightshade, *Solanum nitidibaccatum*, which is notable for its hairiness and also its mottled green berries. Bittersweet, *Solanum dulcamara*, has similar flowers that can be white, though are usually purple, and it is a vine producing red berries. Woolly nightshade, *Solanum mauritianum*, has poisonous berries, but is a tall shrub with hairy grey leaves. Deadly nightshade, *Atropa belladonna*, is not so common in this country and looks quite different, with a red-purple, bell-shaped flower.

Key edible uses

Eaten in many countries, they have a strong liquorice flavour and can be eaten raw or made into jam or chutney but also dehydrate well, becoming a useful garnish on desserts. Fresh-growth leaves (but not older leaves) are widely consumed in Africa and are best blanched and fried, or steamed, and served as a side dish.

Opposite These fresh growth leaves are delicious cooked and served as a side dish. **Inset** White flowers and fully ripe fruit, ready for eating.

Urticaceae

Bed of introduced stinging nettles.

This is the nettle family, which doesn't have many different plants to forage in this country, but nettles (*Urtica* spp.) are such nutritious favourites among foragers that this family is getting separate billing here. Another reason to single it out is that it features the lethal native stinging nettle, which is important to recognise and avoid.

There are four other native nettles, and these are all edible; however, the distinctively small *Urtica aspera* and scrambling, narrow-leaved *Urtica perconfusa* are both at risk, so should be left to sting in peace. The other two natives, the rounded-mound-forming *Urtica australis* and the small and triangular-leaved *Urtica sykesii*, are not declining but are different enough from the introduced species that they can be avoided. Given the quantities of nettles that foragers like to harvest, it is best to stick with the interlopers.

One of these introduced species — the scrub nettle, *Urtica incisa*, with its long spikes of small white flowers — has naturalised but is yet to venture far, so you are currently unlikely to encounter it.

Stinging nettle

SCIENTIFIC NAME
Urtica dioica

OTHER NAMES
European stinging nettle, great stinging nettle, perennial nettle

DESCRIPTION
This is a perennial nettle. Spreading into dense thickets, its vertical stems commonly grow around knee-height but can reach up to 2m. Its spear-shaped serrated leaves are usually deep green and have small needle-sharp hairs that release a stinging chemical when touched. The male and female flowers are produced on different plants. The green-white female flowers are borne in drooping tassels, producing clusters of seeds.

DISTRIBUTION
Throughout drier parts of Aotearoa New Zealand, but more common in the South Island and East Coast regions.

HABITAT
Likes well-drained soils with some nutrient enrichment, often from livestock such as sheep, and also often grows at the base of macrocarpa or pine shelter belts.

SEASONALITY
Available year-round but best harvested in late winter and spring when there tends to be more fresh growth.

FORAGING TIPS
Wear gloves and long trousers and use a sharp long knife or sickle, cutting off the top third of the nettle.

CAUTIONARY NOTES
Forage only new growth as older plants can cause irritation, and be very careful about contact with the leaves: handle with gloves until cooked or dried.

Key edible uses

Rich in iron, magnesium and other minerals as well as various vitamins, nettles need to be cooked or dried to be made safe to eat. But then leaves, stems and roots will make tasty pesto, green sauce, soup, stews, smoothies and stir-fries as well as an excellent side vegetable. The dried leaves and flowers can be steeped to make herbal tea and the shoots can make beer.

Dwarf stinging nettle

SCIENTIFIC NAME
Urtica urens

OTHER NAME
Annual stinging nettle

DESCRIPTION
This is an annual nettle with male and female flowers on the one stem. Growing to only 60cm high, it has deeper serrations than the perennial nettle (page 280) and more-heart-shaped leaves. The deep-green leaves have small needle-sharp hairs that release a stinging chemical when touched. The greenish flowers cluster close to the stem.

DISTRIBUTION
Found throughout Aotearoa New Zealand, particularly in eastern regions.

HABITAT
Likes well-drained soils with some nutrient enrichment such as from sheep pastures, and also found in gardens and under trees.

SEASONALITY
Best harvested in spring.

FORAGING TIPS
Wear gloves and long trousers and use a sharp long knife or sickle, cutting off the top third of the nettle for the fresh growth.

CAUTIONARY NOTES
Forage only new growth as older plants can cause irritation, and be very careful about contact with the leaves: handle with gloves until cooked or dried.

Key edible uses

Rich in iron, magnesium and other minerals as well as various vitamins, these nettles also need to be cooked or dried to be made safe to eat. Once this is done, leaves, stems and roots will make tasty pesto, green sauce, soup, stews, smoothies and stir-fries as well as an excellent side vegetable. The dried leaves and flowers can be steeped to make herbal tea and the shoots can make beer.

Ongaonga ⚠

SCIENTIFIC NAME
Urtica ferox

OTHER NAME
Tree nettle

DESCRIPTION
This is not to be foraged but avoided at all costs. Growing 2–3m high, often in thickets, it has narrow, serrated leaves that grow in pairs and are covered in white needles that can be as long as 6mm. Commonly regarded as Aotearoa New Zealand's most lethal native plant, when touched it releases a toxin that can cause death or at least severe reactions.

DISTRIBUTION
Spread throughout Aotearoa New Zealand from Otago northwards.

HABITAT
Favours edge environments, in coastal and non-alpine forests.

Inset and opposite The poisonous ongaonga even looks scary, with its elongated pointed leaves and long, unfriendly looking needles, which are a key feature of our endemic nettle.

Violaceae

Violets

The familiar faces of viola flowers are easy to identify with their five petals arranged with two pairs on top of each other and a single petal below. The flowers of wild pansies are usually a lot smaller and less vibrant than garden cultivars (*Viola* x *wittrockiana*). The colouring of each species can vary, the purples ranging from blue-purples to violet, mauve and lilac, so treat the following descriptions as generalisations. And to confuse things further, there are other introduced varieties that might start appearing more widely, such as *Viola betonicifolia*, which is making a bid for freedom in Auckland, and the common dog violet, *Viola riviniana*, which is escaping a number of urban areas.

Recent research has shown that wild pansies contain vitamins, antioxidants and various medicinal qualities, proving the old traditions of their healing properties correct.

Wild pansy

SCIENTIFIC NAMES
Viola arvensis, Viola banksii, Viola tricolor, Viola x *williamsii*

OTHER NAMES
European field pansy, heartsease, horned pansy, Johnny jump up, wild pansy

DESCRIPTION
These introduced creeping annual plants have lance-shaped serrated leaves. Usually, *Viola arvensis* flowers are white all over except the bottom petal, which is yellow; *Viola banksia* is white with a purple blotch where all five petals join, leaving white edges to each petal; *Viola tricolor* has purple upper petals, a white or purple bottom petal with a yellow blotch in the centre and dark-purple honeyguides on the lower three petals; and *Viola x williamsii* can be a mixture of any of the above.

DISTRIBUTION
Patchy distribution, mostly found close to urban areas. *Viola banksia* is primarily found in the North Island.

HABITAT
Prefer moist soils and partial shade, often growing near abandoned gardens.

SEASONALITY
Spring to autumn, with the plant often dying away over the winter months.

FORAGING TIPS
Flowers are delicate and best cut off with sharp scissors and placed carefully into a container with a damp paper towel.

SIMILAR SPECIES
There are three native viola that are very similar: *Viola cunninghamii, Viola filicaulis, Viola lyallii*. The flowers are white, though sometimes may be very pale purple. The bottom petal has deep-purple honey guides, which in *Viola filicaulis* might extend to the other petals. This species also has a yellow blob on the lower petal. The leaves of all three are rounded, almost heart shaped, with notched edges. They are not rare so can also be foraged and used in the same way as the wild pansy.

Key edible uses

The flowers are edible raw, providing a distinct aromatic flavour, which in *Viola tricolor* is reminiscent of peas. A great garnish for salads or desserts or for candying or freezing to make colourful ice cubes. The mildly flavoured leaves are also edible, and can be eaten raw in a salad or cooked as a vegetable or used to thicken soups (they contain a small amount of mucilage).

Clockwise from left Wild pansy; white *Viola filicaulis* flower; the tiny heart shaped leaves of *Viola filicaulis* nestled on the forest floor; *Viola cunninghamii* growing on the side of a bank.

Sweet violet

SCIENTIFIC NAME

Viola odorata

DESCRIPTION

A low-growing plant, often growing in dense mats. The heart-shaped, dark-green leaves have scalloped edges, typically 3–5cm across. The distinctively scented flowers are small and usually deep purple (violet).

DISTRIBUTION

Found escaping to the wild close to urban centres, mainly from Christchurch north.

HABITAT

Moist, shaded areas, such as woodland edges, gardens, and damp edges on farmland and river margins. They are also cultivated for ornamental purposes in gardens.

SEASONALITY

Mainly available from June to September; tend to die back in summer.

Opposite All parts of the the sweet violet are edible, from the flower, seed, root, and heart-shaped leaves.

Key edible uses

The sweet-perfumed flowers are particularly good in desserts, for making teas, syrups and candied decorations, or for flavouring honey or vinegar. The flowers and young leaves are mild in flavour and can be eaten raw or cooked and added to salads (savoury and fruit), soups, and other dishes. The leaves can also be used to thicken soups or stews. Leaves, flowers, seeds and roots have been used medicinally and are high in vitamins and beneficial compounds.

Conifers

Catkins of a Monterey pine

Conifers appear in various families but reach back to a group that emerged millions of years ago bearing its seeds on the outside rather than embedded within fruit or nuts. They produce cones, the leaves usually evergreen and like either needles or scales, and the trunks often produce resin. From a forager's point of view they are useful to group together because the cones are a helpful first step in identification and because they are used in a similar way.

Conifers produce male catkin-like cones and different female seed-bearing cones that may be woody, like the pinecone, or look like fleshy berries, as with the tōtara. Depending on the species, the male and female cones may both be on the same tree or only appear on separate trees.

The best conifers worth foraging from in this country are included in this section, but there are others you can check out. It can be confusing trying to work out what family a conifer belongs in, as names such as 'pine', 'spruce' and 'cedar' have been used indiscriminately for members of different families. The conifers covered in this section come from three families: Cupressaceae, the cypresses; Pinaceae, the pines; and Podocarpaceae, the podocarps. However, also be aware of the Phyllocladaceae family of celery pines, whose edible qualities have hardly been explored; the Picea family of spruces and the Abies family of firs, both of which are rare to find in this country; and the only local member of Araucariaceae, the kauri, *Agathis australis*, whose pine-flavoured resin can be chewed like gum. In fact, this was done by early Māori, who picked the resin off kahikatea, too, and you can forage pine resin in a similar way. These sticky resins have a very strong pine taste because of the terpenes they contain; these are safe in small amounts, but avoid high concentrations.

In general, conifers have aromatic flavours that make them useful if made into seasonings or used as a hard herb in the form of a sprig dropped into dishes being cooked. Many conifers are useful for drink infusions, and are included in alcohol such as gin or beer: juniper berries are a staple ingredient in gin; modern distillers are experimenting with all sorts of conifers in their botanical mixes; and Captain Cook made his beer from 'spruce leaves', which are thought to be rimu, mixed with a decoction of mānuka leaves. In addition, spruce and pine needles are high in vitamin C and they make excellent herbal teas.

Of course, conifers don't all taste the same: the fresh-growth tips may have a lemony flavour; some species of the celery pine, such as tānekaha, *Phyllocladus trichomanoides*, are reminiscent of celery or carrot, an under-utilised culinary item thus far; and others have a hint of resin or even turpentine — after all, the latter is a distillation made from various *Pinus* species. Both the unripe and ripe fruits have concentrated flavours that might be aromatic and juniper-like or sweet, as with snow tōtara. And although not fully explored, the pollen and male flowers of podocarps may have some culinary use in the same way that pine pollen does.

Those who are pregnant are advised not to eat conifers at all, especially in the first trimester. And be very careful not to mistake yew (*Taxus baccata*) with any of these species, as its leaves and seeds contain highly poisonous taxine alkaloids (see page 314).

Conifer forms

There are more than 25 species of Cupressaceae, or cypress, in New Zealand, and most aren't considered widely edible. Some cypresses, such as *Thuja plicata*, contain the irritant thujone; others, such as *Cupressus macrocarpa*, contain isocupressic acid which can induce abortions in cattle. When using Cupressaceae plants as a flavouring, only small amounts are advised.

Planted for their wood, Pinaceae, or pines, run rampant across the country. Their clusters of needles and branches grow spirally from the tree. The cones produced are the archetypal form, carrying pine nuts, which are mostly edible though not all species produce large or tasty ones. As a result, only four or five species are cultivated commercially, of which only *Pinus pinea* is grown in this country and you are unlikely to encounter it in the wild. For the forager, though, the pine foliage is the target, being vitamin-rich and having a tender juvenile form before the needles develop. The resin also has a variety of uses, most famously as a flavouring for the Greek wine retsina.

Most of the 24 species of podocarpaceae, or podocarp, conifers in Aotearoa New Zealand are endemic, with only a few introduced species. Mātai, *Prumnopitys taxifolia*, also known as the black pine, has edible purple-black berries that are slimy, sweet and strongly aromatic. It is also a source of gum which is drunk as a bitter beer. As with all native plants, forage with care, ensuring that there is plenty left for birds, you do not endanger the viability of the plant, and you scatter any unwanted seeds where they might grow.

Snow tōtara berries grow inland in alpine basins and at high altitudes.

Lawson cypress

SCIENTIFIC NAME

Chamaecyparis lawsoniana

OTHER NAMES

Ginger pine, Oregon cedar, Port Orford cedar, white cedar

DESCRIPTION

An evergreen tree that can grow up to 50m tall. It has a triangular or columnar shape with dense frond-like foliage composed of scale-like leaves that are flattened and arranged in opposite pairs along the branches. The foliage is greyish-green to bright green and has a distinctive aromatic lemony scent. Its pale-green underside is a useful identification tool: each leaflet tightly packed along the stems is edged in white, creating a diamond pattern over the foliage. The male catkins appear at the end of the fronds and are small bright-red ovals with dark dots; the round female cones are not much bigger than 5mm across and have tiny spikes; they begin blue-green, opening and turning brown and woody as they mature.

DISTRIBUTION

Found throughout Aotearoa New Zealand but absent from higher-altitude areas above 1000m.

HABITAT

Likes areas with well-drained soils but with moderate amounts of rainfall. Often grows in shady aspects with other conifer species in parkland areas or the edge of forestry plantations.

SEASONALITY

Available year-round with more fresh growth in late winter and early spring, and cones through spring and summer.

CAUTIONARY NOTES/CONFUSION SPECIES

Research has found that *Chamaecyparis lawsoniana* does not contain thujone while other cypresses do, so look out for the distinct markings under the foliage detailed above. Do not ingest essential oils made from this plant.

Key edible uses

The fresh foliage contains limonene, the chemical found in citrus peel, and will therefore impart an aromatic lemony flavour when added to soups or other cooked dishes as a hard herb. A tea can also be made from the leaves, or they can be used to infuse syrup or sorbets. Studies have shown that the essential oil has antibacterial and antifungal properties.

Above opposite Lawson cypress has unique catkins that are bright red ovals, while female cones begin blue-green, maturing into brown. **Below** A yellow varietal of lawson cypress.

Japanese cedar

SCIENTIFIC NAME
Cryptomeria japonica

OTHER NAME
Japanese redwood

DESCRIPTION
A large cypress reaching 70m tall. The foliage is green but turns a bronzy-brown over the winter months. Its leaves have a flattened needle appearance and grow outwards from the stems. The male catkins are yellow-green ovals, turning brown, and the female cones begin green and spiky and become brown, rough-looking woody spheres.

DISTRIBUTION
Found throughout Aotearoa New Zealand in ornamental plantings, but more common in the North Island.

HABITAT
Likes areas with a mild to warm air temperature and high rainfall. Most often encountered in areas of parklands and regional parks and in some cases on the edge of forestry blocks.

SEASONALITY
Fresh-growth leaves are best foraged in spring and early summer, with cones foraged in late spring.

FORAGING TIPS
Having a ladder can be useful for accessing cones higher up on the tree.

Key edible uses

A mildly flavoured herb tea can be made from fresh-growth leaves, the tips of which can be lightly steamed. When soft, the green cones can be lightly boiled and eaten. In traditional Japanese cuisine, thin moistened sheets of the wood — sugi — are used as wraps, infusing fish or meat, particularly duck, with its flavours and aroma before cooking. The wood's essential oils have been approved by the US Food and Drug Administration for use as a food preservative.

Inset Japanese cedar catkins are yellow-green ovals, which turn brown as they age. **Opposite** Japanese cedar wood is used as a food wrap in traditional Japanese cuisine.

Larch

SCIENTIFIC NAME
Larix decidua

DESCRIPTION
One of the medium-sized pines, larches often have spindly-structured branches. Unusually for conifers they are deciduous, with foliage from September to June. The foliage is often vivid green, with short needles growing in tight clusters and a small cone. Growing upright and beginning dark pink, the female cone turns brown as it matures. The male catkin is yellow and hangs down.

DISTRIBUTION
An invasive species in Aotearoa New Zealand, larch tends to be found in higher-altitude areas and is more common in the South Island, primarily the Otago, Canterbury and southern Marlborough regions. While rarer in the North Island it can be found in a few sites.

HABITAT
Well-drained soils; particularly likes growing on slopes.

SEASONALITY
Fresh foliage in September to November.

FORAGING TIPS
Having a ladder can be useful for accessing foliage higher up on the tree. Take a lipped bottle to collect the sticky resin. If you're foraging the tree in late spring, keep an eye out for the larch bolete mushroom, *Suillus grevillei* (see page 440), which is another citric-flavoured ingredient worth collecting.

Key edible uses

Fresh-growth leaves can be collected in spring and are good in salads as they have a soft texture. Chopped finely they can be mixed with butter or sugar. The foliage can also be used as an aromatic seasoning with a refreshing lemony flavour. You can infuse the needles for tea or in gin or vinegar, or steep in a hot sugar-and-water syrup for half an hour to make a citric-tasting cordial or a sauce for desserts. The strongly flavoured resin can be collected from a cut in the trunk and dissolved in boiling water for use as a flavouring; only a small amount is required. The young catkins and female buds when purple can be infused, pickled or fermented. The inner bark makes a flour substitute.

Opposite Larch is an invasive species in Aotearoa New Zealand, mainly found in the South island. **Inset** Young, green female cones.

Monterey pine

SCIENTIFIC NAME
Pinus radiata

OTHER NAME
Pine

DESCRIPTION
The straight trunk of this pine produces a pyramidal or columnar shape; it can reach 60m but is typically found at around 5–20m high. The bark of young trees is smooth and grey, while in older trees it becomes thick, rough and deeply furrowed. The leaves are needle-shaped, dark green and grow in bundles of three. The needles are usually 10–15cm long with male catkins small and yellow and female cones large and green, eventually turning brown when mature.

DISTRIBUTION
Found throughout Aotearoa New Zealand in areas of low to moderate rainfall.

HABITAT
Prefers colder-climate areas with well-drained soils.

SEASONALITY
Fresh-growth needles are available year-round. The male catkins are green in winter and early spring, releasing a yellow pollen in spring. Green female cones appear mainly in spring and early summer.

FORAGING TIPS
Use a container to collect pollen by shaking the branches.

CAUTIONARY NOTES
Pinus radiata contains small levels of isocupressic acid, so do not eat if pregnant. The pollen is high in testosterone, which can provide a boost but is another reason to avoid this tree in pregnancy.

Key edible uses

Pine needles can be used for making herbal teas and for flavouring salts, sugars and vinegars. The green catkins are crunchy and can be added to salads or infused or pickled in oil and light vinegar. Pine gum can be used also in the making of an aromatic syrup. Green pinecones can also be infused for making a syrup. Mature pinecones can be heated on a stovetop or in an oven to induce them to open and release their nuts, which you can then rub to remove their winged papery casings. The nuts can then be roasted and used in salads or for pesto. The mild-flavoured pollen can be used as a garnish or as a powder in smoothies and in baking.

Opposite Monterey pines are commonly cultivated for Christmas trees in Aotearoa New Zealand. **Inset** Monterey pinecones forming.

Douglas fir

SCIENTIFIC NAME
Pseudotsuga menziesii

OTHER NAMES
Douglas spruce, Oregon pine

DESCRIPTION
Although it is commonly called a fir and sometimes a spruce, this is a member of the Pinaceae family. A medium to large conifer with orange-scented, dark-green needles that are shorter and softer than pines, and have whitish undersides. Its distinguishing factor is its distinct long female cones, growing at the end of the branches. They have long bracts that protrude from the layers of the cones; initially this female part looks a bit like a pink and yellow bottlebrush, but as it hardens into a cone the bracts become green and feathery, and as the cone matures it becomes beige and hairy-looking.

DISTRIBUTION
Tends to be found in higher-altitude areas and is more common in the South Island.

HABITAT
Well-drained soils; particularly likes growing on slopes.

SEASONALITY
Fresh tips can be harvested year-round but mainly in spring. Green cones are harvested in spring.

CAUTIONARY NOTES/CONFUSION SPECIES
Take care not to confuse with the poisonous yew tree (see page 314).

Key edible uses

The tender new growth at the tips has a distinct citrusy, resinous flavour and is high in vitamin C and electrolytes. It can be eaten fresh, or kept by drying or freezing. The tips will infuse flavour into syrups, cocktails, tea, honey, vinegar or liquors. Alternatively, chopped finely, they can become a seasoning for fish, poultry, baking or desserts, or else used as you would a sprig of rosemary. The sweet citrusy pollen can be collected during spring and used as a natural sweetener or as a seasoning in baked goods, smoothies or dressings. Boiled in sugar-water, the green cones make a lemon-flavoured syrup. The cambium (the layer under the bark) can also be taken from trees and dried out.

Clockwise from left Fresh Douglas fir needles; catkins forming; fresh, green cones; mature pinecone.

Rimu

SCIENTIFIC NAME
Dacrydium cupressinum

OTHER NAMES
Huarangi, puaka, red pine, southern conifer

DESCRIPTION
Rimu is a large coniferous evergreen tree in the Podocarpaceae family, reaching up to 50m tall. It has a straight, cylindrical trunk with a dense, symmetrical crown of dark-green hard-textured leaves, which are needle-like and arranged in spirals around the branches. They are typically about 2–4cm long and have a glossy appearance. The end of their branches begins green but becomes red and swollen, looking like a berry when mature. On the end of this is the blue-black seed. The bark is reddish-brown and has a rough texture.

DISTRIBUTION
This is the most widely distributed conifer in Aotearoa New Zealand, growing from Northland down to Stewart Island. It tends to prefer areas with higher rainfall so has a stronghold on the west coast of the South Island. Generally, it is a low- to mid-altitude species.

HABITAT
Rimu is a characteristic species of both lowland and mountain forests, preferring cool and moist environments, growing alongside other native tree species in mixed bush.

SEASONALITY
Fresh-growth foliage is available throughout the year, fruiting in late summer to early autumn.

FORAGING TIPS
Younger trees provide more-accessible foliage.

CONFUSION SPECIES
The only potential confusion could be with another native, *Manoao colensoi*, the silver pine, which has slightly similar foliage, especially when young; but rimu foliage is much more needle-like and droops in green waterfalls rather than growing upwards.

Key edible uses

The berry is edible and has a distinct aromatic flavour; the seed is usually discarded, which can be fiddly. If eaten in abundance, the berries are reported to cause constipation, but given the height of the trees you are unlikely to gather many. As a result, the fruit is best eaten raw as a garnish. Fresh-foliage sprigs can be used for herbal teas and infusions, providing a mild conifer flavour and a refreshing aftertaste. Traditionally the berries were eaten, while the bark, gum and leaves were used medicinally.

Opposite Both rimu berries and fresh foliage can eaten, but most people don't eat the seed as it can give the unlucky consumer gut issues.

Kahikatea

SCIENTIFIC NAME
Dacrycarpus dacrydioides

OTHER NAMES
Katea, white pine

DESCRIPTION
One of our tallest endemic trees in the Podocarpaceae family, its pointed canopy reaching heights of 65m. Its bark is mottled grey and white, the foliage often thick with tiny leaves forming densely on the spindly branches. Male catkins/pollen form on the end of the foliage. The female fruit — koroī — is orange when ripe and small (diameter less than 5mm), forming in thick green clusters. Often will occur in dense groves, especially in damp soils.

DISTRIBUTION
Found throughout Aotearoa New Zealand but mainly in areas of lower to mid elevation. The last large forests are found in south Westland.

HABITAT
Often in areas with damp fertile soils and around wetlands. Overall likes areas of flat terrain to form forests. Has increasingly been planted in urban areas as part of restoration programmes.

SEASONALITY
Fresh-growth foliage in spring with green fruit from December to February and fruit ripening from March to May.

Key edible uses

The fresh-growth foliage can be used as an aromatic herb flavouring, with sprigs added to stews and soups or steamed hāngī dishes. It imparts an earthy lemony flavour typical of many conifers. Fresh sprigs can also be steeped to make herbal tea or steeped in alcohol to make botanicals. The green fruit are particularly aromatic and can be pickled like capers and used as a garnish. It is generally too hard to collect enough of the small ripe fruit for jam or desserts — and fiddly to remove the seeds — but they can be eaten raw as a garnish or also added to botanical beverages. They are sweet with a slightly bitter aftertaste.

Inset Ripening kahikatea berries in an array of colours. **Above opposite** Ripe kahikatea fruit provides a fresh zing of colour to the forest treetops. **Below** Kahikatea forage is aromatic and herby.

Snow tōtara

SCIENTIFIC NAME
Podocarpus nivalis

MĀORI NAME
Tauhinu (though this name is also given to other very different plants)

OTHER NAME
Mountain tōtara

DESCRIPTION
This endemic member of the Podocarpaceae family is a sprawling, bushy shrub, spreading to several metres in width and height, though often found as a low mat with a bonsai-like appearance. The fleshy lance-shaped leaves spiral up and out from the stalks (rather than being tight like many other podocarps), their fresh growth bright green but dulling to yellowy-brown. The male catkins are like little pinky-red fingers. The female fruit forms on the end of the branches in a similar way to the kahikatea (see page 306), but is a bright pinky-red orb with a green seed protruding at the end.

DISTRIBUTION
Mostly only found inland at high altitudes, generally above 500–1300m, from the Central Plateau southwards, often in alpine basins and slopes.

HABITAT
Sunny alpine environments, generally in rocky areas but also on slopes with well-drained soils. It has been planted in some township areas and is happy in cold climates, as its name suggests, even tolerating snow.

SEASONALITY
Foliage year-round with ripe fruit from mid-January to early April.

FORAGING TIPS
Often you have to tramp long distances to find this species, so take containers to protect the berries. At times, the fruit can be found in prolific amounts.

CAUTIONARY NOTES/CONFUSION SPECIES
It is unlikely to be encountered in the same areas in the wild, but in areas where snow tōtara have been planted take care not to confuse with the yew tree (see page 314).

Key edible uses

The foliage has an aromatic citrus-earthy taste, making it a handy hard herb in cooking and also for infusing alcohol. The green unripe fruit also provides aromatic flavouring, while the sweet floral ripe berries are juicy and make good garnishes or additions to desserts and jams, though the seed has a sharp juniper flavour.

Opposite Snow tōtara tree laden with bright orbs of fruit. **Inset** Snow tōtara berries, with green seeds protruding.

Tōtara

SCIENTIFIC NAME
Podocarpus totara

DESCRIPTION
A large podocarp tree growing to 25m tall. It has a pointed shape when young, progressively becoming more rounded. The spiky leaves grow out from the stem and are short and pointed with a firm texture. The fresh-growth tips are often light green, becoming darker as the foliage matures. The fruit are small and turn from yellow to orange to red with a green seed at the end. Tōtara has a distinct reddish bark that flakes off easily.

DISTRIBUTION
One of our most widespread native trees found throughout the country, with several other species of tōtara growing in alpine zones. It has been planted in many urban environments.

HABITAT
Tōtara favour a wide range of soil types and habitats.

SEASONALITY
Fresh growth occurs in late winter and spring, with ripe fruit on the tree from February to April.

FORAGING TIPS
Look out for areas where kererū are concentrated as they could well be a sign that there is fruit on nearby trees. A ladder may be useful for harvesting higher fruit.

CAUTIONARY NOTES/CONFUSION SPECIES
Take care not to confuse the foliage or fruit with the very similar-looking yew tree (see page 314), which has a poisonous seed.

Key edible uses

When bright and red, the fruit is sweet with a juniper-like aromatic edge. The unripe green fruit can be steeped in aromatic infusions, as can the foliage. Fresh-growth foliage can also be used like a hard herb to add an aromatic earthy flavour to a variety of dishes. Strips of the dropped bark will infuse flavour when wrapped around smoked or steamed ingredients.

Inset Ripe tōtara berries are sweet and aromatic. **Opposite** Tōtara trees can grow to be giants, like the one pictured. The distinct bark flakes off easily.

Miro

SCIENTIFIC NAME

Prumnopitys ferruginea

OTHER NAME

Brown pine

DESCRIPTION

Another podocarp tree that grows up to 25m tall. It has distinct grey-and-white markings on its trunk. The small deep-green leaves are lance-shaped and grow outwards either side of the stems. The male part becomes erect catkin-like cones, while the female produces distinct oval-shaped green fruit that ripens to pink-red and finally a claret colour. The fruit is sized 10–15mm and contains a single seed within.

DISTRIBUTION

Throughout Aotearoa New Zealand but generally in lower-altitude areas particularly with high rainfall, such as the West Coast.

HABITAT

Native forests and occasionally mixed with introduced species. Very widespread and can occur in the understorey of mixed native forests. Also grown as an ornamental native tree in many parkland locations.

SEASONALITY

Fruit from March to May, but the foliage can be used through the year.

FORAGING TIPS

A ladder or picking pole can be useful for getting the higher fruit.

CAUTIONARY NOTES/CONFUSION SPECIES

Take extreme care not to confuse with the poisonous yew tree (see page 314).

The miro might also be confused with its close cousin mataī, *Prumnopitys taxifolia*, also known as the black pine. However, check the ends and undersides of the lance-like leaves: they are sharp and green in miro but rounded and white in mataī. Also mataī produces a purple-black berry, though it is also edible so you may well want to add it to your foraging.

Key edible uses

The berry is edible and has a distinct aromatic flavour and a large seed. Fresh sprigs can be used for herbal teas and flavour infusions, imparting a mild conifer flavour and a refreshing aftertaste.

Opposite Miro, with it deeply green, speared leaves, can be easily confused with the poisonous yew tree. The key difference is actually textural, with yew leaves feeling softer to touch, and miro leaves slightly stiffer. **Inset** Miro drupes are edible, and have a large seed inside.

Yew ⚠

SCIENTIFIC NAME

Taxus baccata

OTHER NAMES

Common yew, English yew, European yew

DESCRIPTION

Yew leaves and seeds contain highly poisonous taxine alkaloids. Its needles are quite short, dark green, flattened and grow on either side of the stems. The undersides of the needles are not whitish as they are in, say, Douglas fir, and they do not have the citrus aroma you'll find in other conifers. The yew produces a bright-red berry, very like a number of native podocarps, and although the berry itself isn't toxic, it carries the seed which can be deadly. Yew toxins are absorbed through consumption, skin contact and inhalation. Even eating a small amount can cause cardiac arrest, leading to death.

DISTRIBUTION

This tree is mainly cultivated, with limited spread into the wild. Often found around inhabited areas, and traditionally planted around cemeteries and parks.

HABITAT

Found throughout New Zealand except in alpine zones.

Opposite Fully ripe yew berries are actually edible, although the seed can be deadly, so it's not worth the risk. Yew toxins are absorbed through eating, skin contact and inhalation, so avoid at all costs.

Ferns

Black tree fern grove

317

There are over 350 species of fern in New Zealand, about 10 of which are recognised for their culinary value. The ferns' koru is usually the primary part of the plant eaten; however, in a few species, such as the hound's tongue fern, the young leaves are consumed.

Care needs to be taken to identify ferns accurately as many species are toxic. In addition, seek out recent resources, as species that were once considered edible, such as bracken, are now considered carcinogenic; future studies may well discover other species to be wary of. Further, bracken and some other fern species contain thiaminase, an enzyme which is harmless if consumed in small quantities or cooked, but large quantities will reduce absorption of vitamin B. The ferns covered in this section are currently recognised as safe (and tasty) to consume when young, but I include them with the caveat that more research needs to be done. Wash all ferns thoroughly after collecting. Avoid eating any fronds that have formed any spore capsules.

Hen and chicken fern

SCIENTIFIC NAME
Asplenium bulbiferum

MĀORI NAMES
Manamana, mauku, mouku, mouki, pikopiko

DESCRIPTION
A medium-sized soft-textured fern with pale-green fronds. They produce bulbils (the miniature leafy plantlets or 'chickens') towards the end of the upper fronds of the 'hen' plant, which will grow into new plants. Similar to the black shield fern, but the pinnules (the subleaflets making up each pinna or leaflet) of the hen and chicken fern are far more finely divided.

DISTRIBUTION
Found throughout Aotearoa New Zealand, but rare and localised on the South Island's east coast, being more common and widespread in wetter regions. Widespread throughout the North Island.

HABITAT
Widespread as an understorey in native and pine forests.

SEASONALITY
Available year-round but best harvested in winter when the koru will be most prevalent and tasty.

FORAGING TIPS
It pays to walk carefully through the ferns to avoid unduly damaging them and to carefully part the leaves to find the koru at the base of the fern clumps. The koru are best removed with a pair of scissors.

CONFUSION SPECIES
Asplenium gracillimum, which is darker green, and *Asplenium* x *lucrosum*, which produces two very different-looking fronds on the one plant. Neither of these will bear many, if any, bulbils. They are unlikely to be harmful.

Key edible uses

Traditionally the leaves were wrapped around eel, birds and kūmara for cooking. They were also eaten as greens and as a relish for potatoes. Young fronds have a bitter taste when eaten raw but are quite palatable when steamed or boiled until just tender. The root can be cooked, but mostly it is the koru that is eaten. These can be preserved in vinegar but are best sautéed or lightly blanched. The koru is mild in flavour but makes a visually striking garnish.

Above opposite This fern produces bulbils, the small shoots protruding from the fronds, that you can see here. These grow into new plants. **Bottom** This bright green fern makes a striking addition to a dinner plate.

Shining spleenwort

SCIENTIFIC NAME
Asplenium oblongifolium

MĀORI NAMES
Huruhuru whenua, pānako, paranako

DESCRIPTION
A glossy, bright-green and medium-sized fern, though its fronds can grow to over 1m in length. Koru are thick-stemmed and fleshy and covered in fine black hairs.

DISTRIBUTION
Widespread throughout coastal parts of the upper South Island and throughout the North Island, extending to inland sites in the upper half of the North Island.

HABITAT
Grows under coastal forest and shrub, often in moist and frost-free environments. Can also occur under old-growth pine forests.

SEASONALITY
Koru are mainly harvested throughout the autumn and winter months.

FORAGING TIPS
Carefully look at the base of the fern for the koru, which are often concealed. Take care to separate out the fronds gently when looking for the koru to minimise damage to the plant.

CONFUSION SPECIES
Similar to *Asplenium obtusatum*, but those ferns have thicker, more fleshy pinnae (or leaves).

Key edible uses

Once the hairs are removed, the young koru can be gently steamed in the same way as asparagus. You can also batter and fry them. They have a crunchy texture and a mild, earthy, asparagus-like flavour.

Inset Glossy shining spleenwort growth
Opposite The koru from this fern can be prepared just like asparagus. Be gentle when harvesting koru so as not damage the plant.

Rereti

SCIENTIFIC NAME

Blechnum chambersii, also known as *Blechnum lanceolatum*, *Austroblechnum lanceolatum* and various other names

OTHER MĀORI NAME

Nini

OTHER NAMES

Lance fern, lance water fern

DESCRIPTION

A small fern that grows in a rough rosette, with oblong pinnae (leaflets) with tips that can be either blunt or pointed. New fronds are commonly pink; mature fronds are green but can have hints of maroon.

DISTRIBUTION

Widespread

HABITAT

Not fussy but often found by streams.

SEASONALITY

Spring and summer.

Key edible uses

Another fern used in the hāngī, though more to cover baskets of food than to be consumed after cooking. Select the young fronds to eat and cook as you would other green vegetables. Traditionally, Australian Aboriginals ate the rhizome roasted or raw.

Inset Veined underside of leaf
Opposite Fresh growth fronds

Kiokio

SCIENTIFIC NAME
Blechnum novae-zelandiae

OTHER MĀORI NAMES
Horokio, koropio, piupiu, rautao, tupari

OTHER NAMES
Palm-leaf fern, water fern

DESCRIPTION
A variably sized fern with fronds from 25cm to 2.5m long. The green fronds often have distinct reddish colouration on the tips, especially in mature specimens.

DISTRIBUTION
Common throughout Aotearoa New Zealand from coastal to subalpine environments.

HABITAT
Often found on forest edges, and common along road cuttings and also roadside water races.

SEASONALITY
Koru are collected in spring and summer.

CAUTIONARY NOTES/CONFUSION SPECIES
There are a number of related ferns, but kiokio can be distinguished from them by the pinnae (leaflets) at the base of its stem, which are very small and rounded. Apart from these, though, the kiokio frond is larger and less spindly than that of many related ferns.

Key edible uses

The fronds were traditionally used for wrapping food for the hāngī. The koru need to be cooked for 10–15 minutes, either simmered on a low heat or steamed. They are best served with a squeeze of lemon and a little salt seasoning. Recent research has found beneficial flavonoids in the fronds.

Inset and opposite Kiokio koru have a distinct red colouration on the tips. The fronds can be used to wrap food for hāngī.

Black tree fern

SCIENTIFIC NAME
Cyathea medullaris

MĀORI NAMES
Mamaku, katātā, kōrau, pītau

DESCRIPTION
A large tree fern with dark-green fronds on both sides, growing off thick black stalks, and tall trunks up to 20m high marked with hexagonal frond scars. The stem has dark-red hairs along its margins.

DISTRIBUTION
The most common and widespread tree fern in Aotearoa New Zealand.

HABITAT
Grows in the damp understorey of both native and introduced forests.

SEASONALITY
Koru are harvested in autumn, winter and early spring.

CAUTIONARY NOTES/CONFUSION SPECIES
Although the pith from the top of the trunk can be eaten, harvesting it will kill the plant, so stick with koru and frond stems. Do not regularly harvest koru from the same plant.

Other tree ferns won't have the dark-red hairs along the margins. Note also the colour of the underside of the leaves, the trunk scars and the height of the trunk (the mamaku is the tallest tree fern in Aotearoa).

Key edible uses

The large koru is harvested before it unfurls and becomes stringy, though the rather tasteless pith of the unfurled stems can also be eaten (traditionally, it was baked in the hāngī for many hours). Focusing on the koru, the dark hairs need to be wiped off, then the koru is best peeled. It can be eaten raw, for instance added to a smoothie, but is usually sliced and boiled, steamed, fried or baked until tender. It can be eaten as a vegetable, added to soup, stewed with fruit, pickled in brine, or used as a substitute for sago or mashed turnip. It is also used as a traditional medicinal poultice. It can be bitter eaten raw, but when cooked has a mild, slightly sweet flavour and is slimy in texture.

Inset Dark red hairs grow along the margins of this koru. **Above opposite** Black tree ferns favour damp forests. **Below** Don't harvest pith from the top of this fern's trunk as it will kill the plant. Stick with koru, and frond stems.

Hound's tongue fern

SCIENTIFIC NAME

Zealandia pustulata, also known as *Phymatosorus pustulatus* and *Lecanopteris pustulata*, and formerly *Microsorum pustulatum* and numerous other names

MĀORI NAMES

Kōwaowao, maratata, pāraharaha, raumanga

OTHER NAME

Kangaroo fern

DESCRIPTION

Has a creeping, thick rhizome (over 4mm in width), producing glossy green ferns with wide green lobes coming off a distinctive dark central stem, though when young the fronds are unlobed, more like a long skinny hound's tongue. Its fronds grow up to 20cm long. The younger fronds are a bright-green colour that dulls as they mature. The fertile fronds have two lines of bulging dark-brown spore-producing capsules on each pinna (leaflet), while the sterile fronds are wider and missing the 'pustules'.

DISTRIBUTION

Found throughout Aotearoa New Zealand.

HABITAT

A very hardy fern, growing in scrub and forests in a wide range of moisture levels and sun aspects and found in various habitats, including on the ground, on logs and creeping up tree trunks.

SEASONALITY

Fresh fronds are available through the year, depending upon local conditions, sprouting following periods of rainfall.

CAUTIONARY NOTES/CONFUSION SPECIES

Similar to *Lecanopteris scandens* (formerly *Microsorum scandens*) or mokimoki, which has a musky scent and duller, narrower pinnae; also to *Lecanopteris novae-zealandiae* (formerly *Microsorum novae-zealandiae*), which also has narrower pinnae, along with orange-brown spore capsules, and is mostly found in the North Island in high-altitude forests. There is also a fern that resembles the juvenile tongue shape, but it does not develop lobes in maturity. Note that the pinnae of the hound's tongue are not divided right at the stem, which distinguishes it from other superficially similar ferns.

Key edible uses

The fresh fronds can be eaten raw in a salad or as a garnish and have a mild pea-like flavour. The leaves can also be lightly steamed. Once fronds mature and darken in colour, they can be used to cover foods that are to be steamed; this will infuse the food with the fern flavour, but discard the fern after cooking as it will be too tough to eat.

Clockwise from left Hound's tongue fern in various growth stages; bright-green growth; this fern likes growing on trees and rocks.

Grasses and Allies

Jointed rush *Apodasmia similis*

Unlike cows, humans cannot digest the cellulose in grass; as well as this, the silica in grasses is the main problem as it grinds away the teeth of animals not adapted to eat grass leaves. However, the seeds are a different matter, and they contain protein, carbohydrates and minerals. Wheat, barley, oats, rye and rice are all different species of grass grain, but there are wild species that can be foraged for free. There isn't a lot of information available on eating wild grass grains, but they are generally considered safe as long as they are not infected by fungi. Animals can have serious adverse reactions to certain grasses, but these reactions are rarely shared by humans.

There is even less information on native species, but I have found many to be edible, including the seeds of the sedge oioi or jointed wire rush, *Apodasmia similis*, which is found across much of the country. The seeds of reeds and sedges (*Juncus* and *Carex* spp.) are mostly bland but will add fibre to flour mixes. And it isn't just the grains — for example, the roots of purua (*Bolboschoenus* spp.) were traditionally eaten by Māori as were the new shoots of the sedge pīngao, *Ficinia spiralis*, but this should be left untouched as sadly it is now in decline because of competition with marram grass, *Ammophila arenaria*. Though feel free to munch on the thin roots of marram grass as they are edible while not exactly sustaining.

The species included here introduce a range of grasses, which mostly require little work. Lightly toasting seeds brings out their nuttier flavours and also dries them, making them easier to grind into flour. It also makes it easier to rub off the chaff. Of course, you may prefer to keep your grain whole to add texture to your dishes: sprinkled over muesli; boiled as you would oats to make a porridge; or mixed with other grains to make seed crackers. And mixing is generally the best way to make the most of wild grass seeds, as they won't necessarily have the depth of flavour of wheat, nor the same nutrients, nor the necessary levels of gluten if you're following standard bread-making recipes. You are also unlikely to forage the quantity needed for many recipes. However, once you start to consider the humble grass, you'll open up a wide range of culinary possibilities.

The largest grass is bamboo, whose new shoots are high in potassium and other nutrients. They have not been included here because, although delicious, they contain a toxin, so require extensive boiling and soaking before they can be consumed. If you forage them, do your research on safe preparation.

Greater quaking grass

SCIENTIFIC NAME
Briza maxima

OTHER NAME
Blowfly grass

DESCRIPTION
A clump-forming grass that typically reaches a height of 30–60cm. It has fine, linear leaves that are usually green or bluish-green. The most distinctive feature is the pendulous flowers, which hang from very delicate stems and are wide, ridged ovals which are cream, sometimes tinged green or purple or brown.

DISTRIBUTION
Found primarily in drier coastal areas of Aotearoa New Zealand, mostly in the northern half of the South Island, the Wellington region and the upper half of the North Island.

HABITAT
Often found in grassy farmland, thriving in well-drained soils and ample sunlight. May be found in coastal habitats, particularly in areas with sandy or rocky soils.

SEASONALITY
The edible seedpods are in season from September to January with a peak in November.

Key edible uses

The seedpods are eaten raw and have a crunchy texture and a mild nutty flavour.

Inset Greater quaking grass has a short foraging season, but the seeds are delicious when you have the opportunity to taste them.

Linseed

SCIENTIFIC NAME
Linum usitatissimum

OTHER NAME
Common flax

DESCRIPTION
An annual herbaceous plant growing to 1m high. It has a slender, upright stem that branches near the top. The 2–4cm-long leaves are alternate, narrow and taper towards the tip. The flowers are typically blue, though can also be white or pink in cultivated varieties. They have five broad petals circled around the central stamens, measuring about 2.5cm across, and are borne in clusters at the tips of the branches. They produce round capsules containing small, flat, oval seeds. These seeds are the valuable part of the plant and are commonly referred to as flaxseeds or linseeds. They are rich in oil and are used in various culinary and industrial applications.

DISTRIBUTION
Mainly found in drier parts of central Aotearoa New Zealand and the upper North Island.

HABITAT
Open and dry edge environments with good sunlight and well-drained soils. Has spread into the wild from cultivated crops.

SEASONALITY
The plant flowers from September to December with ripe seeds from late January to March, and dies back over late autumn.

Key edible uses

The flowers are edible and make an attractive garnish. High in omega-3, the nutritious nutty seeds can be used whole in baking or sprinkled onto various dishes or ground into flour. The flour can go rancid quickly if not stored in airtight containers. It can be used in baking, or as a thickener, as an egg substitute, in place of breadcrumbs or bran, or as a salad sprinkle or added to smoothies.

Inset Linseed is technically not a grass. The seeds are the valuable part of this plant, and add fibre to your diet.

Harakeke

SCIENTIFIC NAME
Phormium tenax (and other species)

OTHER NAMES
Flax, harareka, huruhuruhika, korari, tīhore, wharárika

DESCRIPTION
Harakeke is a robust, evergreen perennial plant that forms large clumps of long sword-like leaves. It can grow up to 1–3m high. The leaves range in colour from green to bronze or reddish-brown, depending on the variety and environmental conditions. The plant produces tall spikes with clusters of tubular red or yellow flowers which become curved seedpods (4–10cm long) containing small dark seeds. The subalpine variety is smaller and has narrower and more-curved seedpods.

DISTRIBUTION
Found throughout New Zealand and one of our most widespread endemic plants.

HABITAT
Uses a wide range of habitats, growing in many edge-zone environments and often planted along roadsides. Naturally occurs in damper areas and especially around wetlands. Can grow in both partial shade and open locations. Often planted in gardens and parkland areas.

SEASONALITY
Flowers appear from September to October, producing fully ripe (jet-black) seeds from January to March.

FORAGING TIPS
Strips from flax leaves can be used for making food ties.

CAUTIONARY NOTES
The leaves are not edible.

Key edible uses

The pollen from flax flowers is a brilliant orange colour and can be collected in the spring; it makes a striking plating garnish. You can also shake sweet nectar from the flowers. The unripe seedpods can be pickled and have a chilli-like heat and a spicy flavour. The fully ripe black seeds have a mild heat and spicy flavour and can be used in salads, added to muesli or used in baking.

Inset The stamens have a stunning orange pollen in the spring which adds a splash of colour to a plate. **Opposite** Harakeke flowers are bright and attractive, with a sweet nectar.

Raupō

SCIENTIFIC NAME
Typha orientalis

OTHER NAMES
Bulrush, cattails, ngāwhā

DESCRIPTION
Raupō grows up to 3m tall, in dense beds or linear thickets. The pointed, strap-shaped leaves are light green in colour and progressively darken to brown as they mature, extending up to 3m long and 2–3cm wide. The white leaf bases have long sheaves, often 30cm or more long, that encircle the stems. The male bulrush has a thick cluster of pollen on the head of the spike and the female a wide cluster of seeds, which are light and fluffy and dispersed by wind when the plant matures. The rhizomes are white inside with a brown outer layer.

DISTRIBUTION
Has a widespread distribution throughout Aotearoa New Zealand as the seeds are easily dispersed in the wind.

HABITAT
Freshwater wetland environments in a wide range of locations other than high-altitude sites. Prefers slow-moving or stagnant waters to grow in, or else the edges of water sources. Generally grows in areas of moderate to high soil fertility and tends to be absent from acidic boggy areas.

SEASONALITY
Fresh shoots in early spring, with the plant growing vigorously through spring and flowering in mid-summer, then dying away from late summer and into autumn.

FORAGING TIPS
Having waders or a kayak to go into the water and cut roots is an advantage.

CAUTIONARY NOTES
Only harvest from areas of good water quality, and rinse plants thoroughly before cooking.

Key edible uses

An important food to Māori, the starchy rhizomes and roots are quite sweet and can be peeled and boiled or roasted and eaten as you would potatoes, or dried and ground into flour for baking. The base of the fresh shoots, harvested in the spring, can be eaten raw like celery or cooked like asparagus. The yellow pollen can be gathered from the flowering bulrush and used in baking, smoothies and also as a colourful plating garnish. If you collect the small seed, it can be roasted to bring out its nutty flavour and used in baking.

Clockwise from left Raupō with pollen; a bed of raupō; new growth shoots; mature seed head.

Succulents, Cacti and Allies

I have grouped species from different families here because the first thing you are likely to notice is the fleshy leaves, which immediately narrows down the identification options. Also, these leaves have developed in response to dry conditions, so you are likely to find plants from these families in similar locations: dry sunny coastal areas with sandy soils.

I have included succulent-like plants in this section, such as purslanes and coastal vegetables, but technically, succulents are a distinct group of plants with very thick, fleshy leaves such as agave and aloe. And as we're talking about dry and sandy habitats, I have also included a cactus which is just starting to become naturalised in a few places around the country. Interestingly, the fruit on both cacti and some succulents are called figs, so perhaps there is a further reason to group them together.

South African ice plant

SCIENTIFIC NAME
Carpobrotus edulis

OTHER NAMES
Ice plant, pig-face fruit, sea fig

DESCRIPTION
A low-growing, spreading succulent with wedge-shaped fleshy leaves 4–10cm long that can vary in size depending on the environmental conditions. It has large flowers (6–10 cm across) with multiple petals radiating around a central disc and a range of colours: pale or bright-pink, purple, or white. Once it has flowered, the plant forms a bulbous fruit with spiky points, which turns from green to yellow as it ripens, and when fully seeded is brown.

DISTRIBUTION
Found throughout Aotearoa New Zealand. as an invasive weed, but largely absent from the west coast of the South Island

HABITAT
Mainly found in coastal areas, sand dunes, cliffs, sloping banks and disturbed habitats, due to its ability to tolerate salt spray and poor soil conditions. Primarily coastal in distribution but can occur at inland environments, often growing on exposed clay cliffs.

SEASONALITY
Available year-round, the fresh leaves and flowers appear in late spring and early summer, with ripe fruit able to be harvested from mid-summer to mid-autumn.

FORAGING TIPS
Use a sharp knife to collect the fruit. Often used for sand dune stabilisation, so take care not to unduly disturb sensitive sand dune environments.

CAUTIONARY NOTES
Large quantities of this plant can act as a laxative.

SIMILAR SPECIES
The Chilean ice plant, *Carpobrotus chilensis*, is similar to this variety, though has smaller leaves, fruit and flowers. The leaves are usually bluey-green, with pink-purple flowers, radiating around a central disc, also producing a small fruit. It can be used in many similar ways to the South African ice plant.

Key edible uses

The fruits, also known as sea-figs or pig-face fruit, have a tangy, tart flavour, are fleshy and are filled with juicy pulp. They can be eaten raw or used to make jam, jellies and desserts or added to salads. Fully ripe fruits are also suited to pickling. The leaves are astringent when raw but can be pickled in vinegar. Flower petals can also be added to salads and have a mild flavour.

Clockwise from left Freshly picked South African ice plant petals; South African ice plant fruit; Chilean ice plant in flower.

Miner's lettuce

SCIENTIFIC NAME
Claytonia perfoliata

OTHER NAMES
Cuban spinach, Indian lettuce, winter purslane

DESCRIPTION
Not so obviously a succulent, especially when young. When fully formed, its bright-green leaves are circular to heart-shaped with several tiny five-petalled white flowers in the centre. Grows in dense beds.

DISTRIBUTION
Throughout Aotearoa New Zealand but tends to be most common around urban centres, and can occur well inland in some Central Otago locations.

HABITAT
Prefers sheltered and shady areas in moist but well-drained soils, and is often found in edge environments.

SEASONALITY
An annual species with fresh growth in mid-winter and fully formed leaves by October. Often dies away over late spring. Out of season in summer, but small amounts can be found year-round.

FORAGING TIPS
A delicate plant, so place in containers when foraged to prevent it from getting crushed.

Key edible uses

A very mild spinach-flavoured green with a crunchy texture. Ideal in salads and as a base for pesto, though can be cooked and eaten like spinach. Can also be made into a green juice. Its bright colour and angular leaf shape make it a desirable plating garnish. Contains oxalic acid, see page 12.

Native ice plant

SCIENTIFIC NAME

Disphyma australe

MĀORI NAMES

Horokaka, ngarangara, ruerueke

DESCRIPTION

Growing in clusters as a groundcover, this indigenous ice plant has a small dark-green wedge leaf 1–4cm long with a pointed tip. The flowers are white or light pink or purple, and the fruit is red. When leaf size is greater than 4cm it is likely that the ice plant is either a hybrid with an introduced species or one of the introduced ice plants.

DISTRIBUTION

Throughout Aotearoa New Zealand but with a patchy distribution in coastal areas. Rarely found away from the immediate coastal zone.

HABITAT

Likes to scramble over rocky walls, cliffs and steep rocks, and coastal slopes. Occasionally also grows on clay banks or as a flat groundcover.

SEASONALITY

Available year-round, flowering primarily in the spring.

FORAGING TIPS

Often grows in small patches, so harvest with care to allow sustainability. When harvesting, cut the outer growth and avoid disturbing the roots, to allow regrowth.

CAUTIONARY NOTES

There is a Chatham Island ice plant (*Disphyma papillatum*) which is at risk, so do not harvest it: it has much smaller leaves.

Key edible uses

The leaves are edible raw and often have a salty umami flavour. When growing in dry conditions, the flavour often becomes intensified and can have an appealing minerality. Ideal for adding to salads, good to cut for a salsa mix or to use as a garnish for seafood dishes. The flower is also edible with a mild flavour. The fruit is too small to be of any notable use.

Above opposite Native ice plant leaves are a dark green wedge, with a salty flavour. **Below** This plant likes growing on rock faces in mats.

Jellybean ice plant

SCIENTIFIC NAME

Disphyma crassifolium; also a similar subspecies, *Disphyma crassifolium* subsp. *clavellatum*

OTHER NAMES

Purple dewplant, round-leaved pig-face

DESCRIPTION

Forming dense mats, the small leaves (1–3cm long) are thick, cylindrical, and range in colour from green to reddish-purple depending on the amount of sun exposure. The flowers are a bright pink-purple. The plant forms a small indistinct fruit.

DISTRIBUTION

Found at coastal locations in Otago and Southland and a few sites in urban centres such as Auckland, Wellington and Nelson. A species which is found in small amounts at the moment but is likely to increase its range.

HABITAT

Thrives in sandy or rocky soils in coastal environments, often on rock walls or exposed banks, or the edges of salt-marshes.

SEASONALITY

Available year-round with flowers primarily in the spring.

FORAGING TIPS

Grows in dense mats, so take the outer leaves to minimise disturbance to the plant's root structure. As it grows low on the ground, wash thoroughly in fresh water before eating.

Key edible uses

The leaves have an intense, tangy, salty flavour and can be eaten raw; especially tasty when added to salads or used as a garnish for seafoods. The flowers are also edible and have a mild flavour.

Prickly pear

SCIENTIFIC NAMES

Opuntia ficus-indica (Indian fig), *Opuntia monacantha* (drooping prickly pear), *Opuntia robusta* (wheel cactus)

DESCRIPTION

These are not succulents but cacti, growing up to 4m high but usually a lot less, on thick trunks with large, grey-green, fleshy, paddle-like stem sections, which have a few well-spaced spines and yellow flowers on top. The yellow flowers on Indian fig produce fruit that is yellow to red, looking a little like a colourful fig. Drooping prickly pear can, as its name suggests, be drooping in form; it has fewer spines and its yellow flowers are marked with red, while its fruit, like that of wheel cactus, is red-purple. Wheel cactus has the largest paddles of the three and they are much more circular.

DISTRIBUTION

Drooping prickly pear is more widespread, appearing patchily in mostly coastal areas close to urban centres across the North Island and in Christchurch and Nelson in the South Island. Indian fig appears only in a few urban centres, and wheel cactus currently only in the Christchurch region.

HABITAT

Recently naturalised, they can be found in sandy beach areas, close to settlements.

SEASONALITY

Flowering in autumn, followed by fruit.

FORAGING TIPS

Wear thick gloves when you forage for cactus, as the spines are sharp.

Key edible uses

The Indian fig is the cactus most widely consumed, but the others are also edible. Remove the prickles from the fruit and the paddles before consuming. The fleshy, watery paddles taste of green beans and are sliced (remove the skin if you prefer) and often served fried with egg and chillies. They and the fruit can be eaten raw or cooked — fried or boiled — or juiced. The fruit, which is sweet and very high in vitamin C, can also be used to make jam and liqueur, and can be dried. The seeds can be ground into flour.

Purslane

SCIENTIFIC NAME
Portulaca oleracea

OTHER NAME
Little hogweed

DESCRIPTION
A ground-creeping plant with thick red stems that radiate out, bearing small leaves which are elongated green ovals, sometimes with red edges. It has a yellow flower, a little like a tiny buttercup except that the five petals are heart-shaped. They produce capsules containing fine black seeds.

DISTRIBUTION
Found throughout the country, generally in coastal and low-altitude areas with low rainfall. Mainly from Christchurch northwards.

HABITAT
Likes well-drained soils and relatively dry environments. Often found growing around concrete and along shingle patches and stony open ground.

SEASONALITY
An annual species that grows from December until May and dies away over the winter months. The peak season is from January to March.

CAUTIONARY NOTES
Take care to wash well and avoid collecting from areas with lots of cars or where the soil may be polluted.

Key edible uses

High in nutrients, the whole plant is edible. The salty lemon leaves and flower buds are eaten raw as a garnish and in salads, or the leaves cooked and used to thicken soups and stews. The stems can be juiced, and the seeds used whole or ground to add to flour or seed mixes. Contains oxalic acid, see page 12.

Beach spinach

SCIENTIFIC NAME
Tetragonia implexicoma

MĀORI NAMES
Kōkihi, rengamutu, rengarenga, tūtae ikamoana

OTHER NAMES
Tetragonia trigyna, bower spinach, native spinach

DESCRIPTION
Grows as a dense groundcover but can also creep up onto dead and live structures. The young stems are purple and the fleshy leaves are typically arrow-shaped, but can be oval or strappy. They are typically bright green but can also be reddish (when heat-stressed) with a frosted colouration on the underside. They produce tiny yellow flowers with four narrow petals and a reddish-purple berry that has a large stone.

DISTRIBUTION
Only patchily distributed in coastal areas.

HABITAT
Usually in coastal areas, but can grow several kilometres inland and up to about 100m altitude. Likes well-drained soils. Often found growing under pine and macrocarpa forests or under coastal shrubs with twigs and branches on the ground that the plant can use as anchor points to grow over.

SEASONALITY
Available year-round but often in best condition in winter and early spring for fresh-growth leaves. Berries available over summer and early autumn.

FORAGING TIPS
Often grows in dense mats. Harvest the leaves and outer stems without damaging the roots, to allow regrowth. When harvested regularly at the same location, you can get a uniform regrowth of fresh leaves and some harvesting tends to promote rapid regrowth. With good rainfall the same plant can be re-harvested at about three-weekly intervals.

CAUTIONARY NOTES
Take care not to use in large amounts over a prolonged period of time, as beach spinach is high in oxalic acid.

Key edible uses

With a salty, savoury flavour, the young leaves are tasty eaten raw. Larger leaves are more bitter and so are suited to cooking or juicing, used as a replacement for spinach or blended into pesto. The red berries can also be eaten but have a large stone. The red stems can be juiced.

Opposite Beach spinach grows in mats, and can creep up onto tree trunks. **Above inset** Ripe berries are scarlet red. **Below inset** This plant's leaves are delicious raw, when young.

New Zealand spinach

SCIENTIFIC NAME

Tetragonia tetragonioides

MĀORI NAMES

Kōkihi, rengamutu, rengarenga, tūtae ikamoana

DESCRIPTION

While thick-leaved, this low-growing, mat-forming plant is less like a succulent than beach spinach, *Tetragonia implexicoma* (see page 350). It has chunky stems that run along the ground, with side stems that grow up to 50cm long with large arrow-shaped leaves (up to 10cm long) with a distinct frosted appearance on the underside. The plant has a small yellow flower which has four petals but they are much broader than in *Tetragonia implexicoma*.

DISTRIBUTION

A patchy, primarily coastal distribution throughout Aotearoa New Zealand.

HABITAT

Often found just above the high-tide mark along the coastline in stony or well-drained and often nutrient-enriched soils.

SEASONALITY

Available year-round.

FORAGING TIPS

Pick only a part of each patch, to allow the plant to regenerate. Increasingly becoming at risk in the wild as introduced pests, especially rabbits, have reduced its range and abundance. However, its seeds are readily available from commercial seed suppliers as it is becoming a popular species to grow as an edible groundcover.

Key edible uses

The leaves have a beautiful crunchy texture and a salty, savoury flavour. A good spinach to use fresh in a salad or as a plating garnish. When cooked, the leaves go deep green and have a very pleasant savoury flavour. Both the leaves and the stems are suitable for juicing for smoothies. Overall, a tastier but more vulnerable plant than the related beach spinach. In some locations the species have hybridised. Contains oxalic acid, see page 12.

Inset New Zealand spinach leaves are triangular, with a frosted appearance on the underside.

Wall pennywort

SCIENTIFIC NAME
Umbilicus rupestris

OTHER NAME
Navelwort

DESCRIPTION
It has succulent, circular, serrated leaves with a central depression, and these are arranged in a basal rosette. The plant produces spires of small bell-shaped flowers, which vary in colour from a pale greeny-pink to light purple.

DISTRIBUTION
Mainly coastal, found around urban centres such as Dunedin, Christchurch and Wellington. Appears to be absent from the upper North Island.

HABITAT
Likes well-drained soils and often grows in sheltered rocky environments and crevices.

SEASONALITY
Found year-round but often dies away over the summer months.

FORAGING TIPS
Forage only fresh growth, as the older leaves are less palatable.

Inset Wall pennywort leaves have a unique central depression.

Key edible uses

The young leaves have a distinct pea flavour and can be eaten raw as a salad green.

Other Flowering Greens

Mullein in flower

You often don't have to go further than your garden path to find some unexpected edibles. Daylilies, for example, are mostly edible, the species *Hemerocallis citrina* being used in Chinese cuisine. Many *Campanula* species are edible, as are chrysanthemum, carnation and gladioli flowers. The petals of sunflowers are edible as well as its nutritious seeds, and its flower buds can be cooked like an artichoke.

Conversely, there are also poisonous plants lurking in your flowerbed, such as the particularly lethal monkshood, *Aconitum napellus*, Italian arum, *Arum italicum*, larkspur, *Delphinium* spp., foxglove, *Digitalis purpurea*, caper spurge, *Euphorbia lathyrus*, swan plants, *Gomphocarpus* spp., bluebells, *Hyacinthoides* spp., the berries of lantana, also known as verbena, *Lantana camara* var. *aculeata*, daffodils, *Narcissus* spp., inkweed, *Phytolacca octandra*, buttercups, *Ranunculus* spp., the tropical periwinkle, *Vinca rosea*, and arum lilies, *Zantedeschia aethiopica*, among others.

The rule is to focus on a certain species, learn how to identify it, be aware of what it might be confused with and, once mastered, slowly add to your repertoire. I am constantly learning.

Onionweed

SCIENTIFIC NAME
Allium triquetrum

OTHER NAMES
Three-cornered leek, wild garlic (mistakenly, as this refers to a different plant)

DESCRIPTION
Growing from small white bulbs, it has strappy, grass-like leaves and produces white bell flowers with pointed petals and generally a green stripe down the centre of the inside of the petals. The Latin name refers to its distinctive three-cornered flower stalks. It has a strong onion smell, especially when you crush the leaves. The whole plant is edible, making a good alternative to spring onions. Regarded as invasive, harvesting is guilt-free.

DISTRIBUTION
Patchy distribution across the country, but grows in abundance.

HABITAT
Common throughout Aotearoa New Zealand, often under trees and on path edges. Prefers moist, fertile soils in sheltered, frost-free locations.

SEASONALITY
Leaves appear from June to December with flowering and peak growth period through spring.

CAUTIONARY NOTES
The onion smell will help you distinguish the plant from others. Do not confuse with white bluebells (*Hyacinthoides* spp.), which do not have the three-cornered stalk. All parts of bluebells are toxic. White petals with green dots are snowdrops *Galanthus nivalis*, while lily of the valley, *Convallaria majalis*, is shorter, with broad leaves, and the rounded-bell flowers have slightly upturned tips. Both of these plants are toxic.

Key edible uses

This plant makes an excellent substitute for spring onions. The leaves, flowers and bulbs can be eaten fresh in salad, salsa verde or pesto and chopped in herb butter. The leaves can also be sautéed with lemon or added to stir-fries, soups and stews. The flowers can make tempura; the bulbs can be steamed or lightly pickled, like cocktail onions, as can the pods to make alternatives to capers.

Above opposite Onionweed flowers are bell-like with a three-cornered stalk — be careful not to mistake them for the toxic white bluebells. **Below** Freshly picked leaves make a great substitution for spring onion.

California poppy

SCIENTIFIC NAME
Eschscholzia californica

OTHER NAME
Golden poppy

DESCRIPTION
California poppy grows up to 30cm high. The long, greyish-green leaves are divided into fine, lobed segments, giving them a feathery appearance. Each stem holds a single bright-orange-yellow flower, which has four petals forming a cup around central golden stamens. A capsule develops holding the seeds.

DISTRIBUTION
Found primarily in drier eastern and central parts of Aotearoa New Zealand. Largely absent from wetter regions and high-altitude environments.

HABITAT
Likes well-drained soils and sunny open aspects, often found along gravel riverbeds and edge environments.

SEASONALITY
Flowers from September to December with the seeds developing over summer. Tends to die back in many areas over the winter months.

CAUTIONARY NOTES
The plant has mild sedative properties so should not be consumed in large amounts. While not the same species it is in the same family as the other type of poppy (*Papaver* spp.), which should not be consumed as it contains toxic alkaloid compounds in stems, leaves, flowers and seeds. Most *Papaver* in this country are red, but there are a few orange or yellow Welsh poppies, *Papaver cambricum*, growing in the South Island. Unlike the California poppy, *Papaver* petals tend to have a crinkly appearance and their leaves are not so fine and feathery.

Key edible uses

The flower petals and leaves can be used as a garnish and mixed into salads, or used as a decorative feature when baking. The seeds are also edible when roasted, or used in cereal mixes or baking. The roots can also be eaten.

Tradescantia

SCIENTIFIC NAME

Tradescantia fluminensis

OTHER NAMES

Inch plant, small-leaf spiderwort, speedy Henry, wandering Jew, wandering willie

DESCRIPTION

A spreading plant forming thick carpets with its fleshy, trailing stems and glossy, dark-green, oval leaves, which are pointed and can sometimes have purply undersides. It produces small white three-petalled flowers.

DISTRIBUTION

An invasive weed in coastal South Island urban centres, particularly from Christchurch north, and across the North Island.

HABITAT

Prefers moist soils in shady areas and frost-free regions. Often found in edge environments, invading bush margins and wastelands.

SEASONALITY

Year-round foliage, flowers in summer.

FORAGING TIPS

Eat soon after picking as the leaves do not keep long.

Inset Tradescantia flowers are delicate, with three petals. **Opposite** You'll probably recognise this invasive weed if you've ever had an overgrown garden! Try the leaves next time you're out for a spot of weed-whacking.

CAUTIONARY NOTES

The juice of the plant may cause skin irritation in some people. Also consume in moderation as the plant contains oxalate crystals, which can irritate the tongue and throat.

Key edible uses

With potentially medicinal qualities, the blandly flavoured leaves can be included raw in salads or smoothies or else eaten cooked: steamed, boiled or added to stir-fries, soups or stews.

Common mullein

SCIENTIFIC NAME
Verbascum thapsus

OTHER NAMES
Aaron's rod, torches

DESCRIPTION
A distinctive plant with large, greyish-green leaves growing up to 40cm long with a grey, furry covering. The leaves radiate around a central point from which a tall — up to 2m high — leaf-covered stem arises that becomes the flower spike. When mature, the flowers are yellow and they may arise singly or in groups. The seeds sitting on the stem where the flowers previously clustered are brown.

DISTRIBUTION
Throughout drier parts of Aotearoa New Zealand, though a cooler-climate species.

HABITAT
Open areas of dry farmland, rough pasture and braided river margins with well-drained soils.

SEASONALITY
All year round.

CAUTIONARY NOTES
While no side effects have been reported, a study has found that the leaves, in particular, contain some quantities of rotenone (toxic) and coumarin (an anticoagulant), so while further research is required it is advisable not to consume this plant other than as a tea. Furthermore, the hairs on the leaves can cause irritation, so eating them is not advised.

Key edible uses

Used in traditional medicine, the leaves and flowers are best dried to make herbal tea.

Other vines

Fruit-salad plant

This section brings together plants from different families, but it makes sense to do so given that their climbing habit is the first thing we notice about them. Do, though, check out the Passifloraceae section on pages 202–209, which covers the range of different passionfruit species that have also wound their way around the country.

So too have members of the trumpet-flowered Convolvulaceae family, also known as bindweed, which includes *Calystegia* and *Convolvulus* species, as well as the particularly invasive blue morning glory, *Ipomoea indica*. Although the new shoots, leaves, seeds and rhizomes of some varieties have traditionally been eaten in different parts of the world — indeed, our own beloved kūmara (*Ipomoea batatas*) is related to morning glory — it is looking increasingly likely that many varieties are not lethal but at the same time not safe to consume. *Ipomoea indica* is regarded as toxic to humans, particularly the seeds; a feature shared with kūmara seeds. Furthermore, recent research has found that many other Convolvulaceae species contain worrying levels of toxic alkaloids. Given these findings, it is best not to forage this family — and in addition, there are some native species that include several that are at risk.

The changing view on bindweed is a reminder to keep on top of research. New findings are being made all the time; and bear in mind that many plants have never been fully tested. Meanwhile, there are still plenty of vines to forage instead.

Kiekie

SCIENTIFIC NAME

Freycinetia banksii

OTHER NAMES

Freycinetia baueriana, pēia

DESCRIPTION

An evergreen, woody, thick-stemmed climbing plant with long, strap-like leathery leaves that are often dark green and glossy and can grow up to 1.5m long, looking a little like a young cabbage tree. It has aerial roots and can be seen on the forest floor, but mostly climbs tree trunks. Forming in the centre of a leaf clump, the large angular flowers are cream-coloured and very ornate, typically consisting of three large triangular petals. Growing up to 15cm long, the bobbly fruit are green oblongs, turning yellow then brown as they ripen.

DISTRIBUTION

Found in forests, particularly in coastal lowlands in the upper west coast of the South Island, the Wellington region, and the west and upper half of the North Island.

HABITAT

Happy in lowland to montane forests, growing on a range of soil types.

SEASONALITY

Flowers are foraged in spring, with ripe fruit through summer and early autumn.

FORAGING TIPS

The fruit is sweetest when ripe, though both the flowers and the fruit are rarely encountered nowadays in many areas as they are a favoured food for possums, so look for an area with some predator control.

Key edible uses

Both the white bracts around the flowers and the fruit can be eaten raw. The bracts have a sweet pear flavour, and can be used fresh in salads, covered in tempura and fried, used to make jelly or as a garnish for desserts, or infused into teas, syrups, vinegar and oils to add a unique fruity essence. The bracts can also be used to make a very aromatic alcohol infusion. The fruit needs to be peeled of its bitter, bumpy outer skin to reach the slightly less bitter, juicy pulp, which is usually eaten raw. Fresh-growth shoots are tender and have a mild nutty flavour and can also be eaten raw, or cooked. The sap can be consumed as a source of water when collected by cutting the stem of the plant and allowing the liquid to drip out.

Opposite Kiekie flowers, and fruit, are a rare find in the wild these days as possums love them. Forage in areas with predator control.

Hops

SCIENTIFIC NAME
Humulus lupulus

DESCRIPTION
A deciduous vine creeping up to 10m high. Its leaves generally have serrated leaves with three to five deep lobes and well-defined veins. The male flowers are small and greeny-yellow, while the female are pale green and cone-shaped, droop off the stems in clusters and hold the fruit. It is the female flowers that are used as 'hops'.

DISTRIBUTION
Found throughout Aotearoa New Zealand, especially in lowland temperate areas with good rainfall, though mostly absent from Auckland northwards.

HABITAT
Grows over scrub in edge zones, especially along river margins, abandoned gardens and sheltered areas with rank shrub. Usually they are found near areas of habitation but can spread into remoter sites.

SEASONALITY
Fresh shoots in spring with the green seedheads (hops cones) collected in late summer. The vine dies away over the winter months.

FORAGING TIPS
A picking pole can be useful for harvesting ripe hops.

CAUTIONARY NOTES
Some people can experience skin irritation when handling sticky hops strobiles, so proceed with caution.

Key edible uses

Fresh-growth shoots have a crunchy texture and can be steamed with the young leaves, which can also be added raw to salads. Both leaves and hops can be dried to make tea. The fragrant hops are bitter and primarily used for flavouring beer, but offer a wide range of applications as an aromatic seasoning, imparting a refreshing flavour to cooked dishes.

Inset Hops are actually the female flowers of this vine.

Japanese honeysuckle

SCIENTIFIC NAME

Lonicera japonica

DESCRIPTION

This is a woody vine that can grow up to 9m long on stems that are red when young. It has opposite, oval-shaped leaves that are dark green and often glossy. The vine produces tubular flowers that open into two tongues, the top one broad and the bottom one narrow, which are white or yellowish-white and turn pale yellow as they age. They produce small black berries. One of the most distinctive features of Japanese honeysuckle is its sweetly fragrant flowers.

DISTRIBUTION

Very widespread as an invasive plant throughout Aotearoa New Zealand but absent from drier areas and alpine zones.

HABITAT

Often thriving at the edges of woodlands and stream banks, the vine reaches for sunlight by climbing over other vegetation. It is known for its ability to escape gardens and colonise disturbed areas, including roadsides, abandoned fields and construction sites.

SEASONALITY

Grows and flowers throughout the year in many locations.

CAUTIONARY NOTES

The leaves contains low amounts of saponins, so should not be consumed in large amounts or regularly over a long period of time.

Key edible uses

The leaves should be cooked, such as boiled or steamed, to be used as a side green and are used in Asian cuisine. The flowers are edible and have a sweet nectar useful for infusing into syrups or desserts. The leaves, flowers and fresh buds are used in herbal tea.

Fruit-salad plant

SCIENTIFIC NAME
Monstera deliciosa

OTHER NAME
Swiss cheese plant

DESCRIPTION
Although it can grow up to 20m high in the wild, it is more commonly found in this country scrambling to 10m. The huge, glossy, heart-shaped leaves are deeply indented and often pierced with holes. The vine produces a white hood around a yellow spire about 15cm long, which develops into a slightly larger, green cylindrical fruit with a patterned texture of scales, looking a bit like a closed pinecone. Over a year or so, the fruit turns yellow and the scales drop off as it matures, producing a sweet tropical aroma, like a fruit salad.

DISTRIBUTION
Found close to urban centres, mostly from Tauranga north as the vine is not frost-hardy.

HABITAT
Prefers moist, rich soil in humid shady areas, such as the bases of banks or trees, from where it grows up towards the sun.

SEASONALITY
The fruit commonly spends most of a year on the vine before it has ripened sufficiently to eat.

FORAGING TIPS
As long as the fruit has started to turn yellow and has started to release its aroma, you can harvest it and place it in a paper bag until the scales have dropped off. You can scoop out the pulp where the scales have fallen, but wait for the remaining scales to drop in their own time.

CAUTIONARY NOTES
Consume only when ripe as it has a needle-like texture when still immature, which is uncomfortable to eat and may be toxic. The sap from the stems can cause skin irritation.

Key edible uses

High in vitamin C and flavoured like an array of different tropical fruits, the sweet white pulp is eaten raw or cooked into desserts or jams. Make sure the fruit is fully ripe, that the scales have dropped and that you remove any black specks before consumption.

Opposite Fruit-salad plant is commonly known by part of its scientific name, *Monstera*. It's grown in popularity as a house plant recently, but can found in warmer urban centres.

Supplejack

SCIENTIFIC NAME

Ripogonum scandens

MĀORI NAMES

Akapirita, kareao, karewao, kekereao, pirita, taiore

DESCRIPTION

A creeping vine that spawls through the forest and coils up trees. Although woody and its lower parts present as a tangle of bare dark vines, its upper parts are soft (hence the common name) and have bright-green medium-sized oval leaves with pointed tips and distinct veins. The clusters of fruit are green, turning red when ripe.

DISTRIBUTION

Found throughout Aotearoa New Zealand, usually in lowland and coastal locations, but more widely in the upper half of the North Island.

HABITAT

Grows in both native bush and forest, defining part of the understorey. Tends to favour frost-free lower slopes and valley floors.

SEASONALITY

Fresh shoots are more common in spring but can be found year-round. The vine produces ripe berries over the summer.

FORAGING TIPS

Generally, the fresh-growth tips are found in patches of native bush that have some predator control as they are sought out by deer and possums as well.

Key edible uses

When still tender, the fresh-growth tips emerging from the forest floor are the key part of the supplejack vine that is consumed. While they can be eaten when raw and crunchy, the flavour is enhanced by light steaming. The root can also be harvested and eaten. The ripe red berries are also edible, but less accessible.

Above opposite Supplejack berries in various stages of ripeness. **Below** The twisted vine of the supplejack is a distinctive feature in the understory of native forests.

Choko

SCIENTIFIC NAME
Sechium edule

OTHER NAMES
Chayote, mirliton

DESCRIPTION
A member of the gourd family, this vine has tendrils that it uses to grow up whatever support it can find, sometimes reaching 10m or more. It has very large leaves, which are an angled heart shape, similar to the leaves of cucumber or grapes. The flowers are relatively small, pale-yellow, five-petalled stars and produce large, pale-green, pear-shaped fruit with slight ridges and sometimes prickles, growing around a single large seed.

DISTRIBUTION
Confined mainly to urban centres, mostly in the North Island, it is particularly common in the Auckland region.

HABITAT
Found scrambling out of gardens into wastelands and edge environments.

SEASONALITY
Fruits usually through autumn and winter.

CONFUSION SPECIES
Do not confuse with moth plant, *Araujia sericifera*, which is toxic, particularly the sap, which can cause skin irritation. It produces a similar fruit to choko, but this is soft and light and contains silky hairs and multiple seeds; also its leaves are small ovals with felty undersides, its flowers are white, pale pink or purple, and it has a milky sap.

Key edible uses

High in vitamin C and minerals, the crispy, rather bland, zucchini-flavoured fruit is excellent combined with spicy sauces or strongly flavoured ingredients. Younger, smaller fruit can be eaten unpeeled: finely sliced or grated raw into salad, especially in combination with a citrus or salad dressing; or cooked and eaten like pumpkin — baked, boiled, fried, breaded, mashed, steamed, stuffed or added to stews or soups. It can also be pickled in vinegar. The rest of the plant is also edible: the tubers treated like root vegetables, the vine tips and leaves chopped into salad, pesto, stir-fries or soups. The seed is also nutritious; it has a nutty flavour and can be eaten raw when small, young and relatively soft.

Opposite Choko looks very similar to the toxic moth plant — don't get them confused! **Inset** Choko has a variety of uses, and can be eaten raw, cooked, and pickled.

Other Trees and Shrubs

Bay tree

When you start to identify a plant, it is reasonable to begin with the most obvious elements and then work down to the small details. But those obvious elements are not always so clear-cut — for instance, how do you determine whether a plant is a large shrub or a small tree? One might encounter a garden fig, *Ficus carica*, out in the wild and, seeing its smaller form and multiple branches close to the ground, search among shrub species to confirm its identification. And yet figs are generally referred to as trees, which is understandable when you think of the tall single trunk that only branches high off the ground in the Moreton Bay fig, *Ficus macrophylla*.

Once you establish the connection, though, the relevant question for the forager isn't whether it is a shrub or a tree, but whether other *Ficus* fruit are as edible as those you'll encounter in the garden. The answer is that most figs produced by other *Ficus* are indeed edible, just not as tasty. The small fruit produced by the huge Moreton Bay fig, for example, are edible but quite dry and seedy. The same goes for Port Jackson figs, *Ficus rubiginosa*, which are popping up in some areas from Auckland northwards, and likewise you can sample the fruit of the climbing fig, *Ficus pumila*; but it's the garden escapee that is really the one worth foraging. There are a few *Ficus* varieties with sour fruit that could give you stomach ache but you are unlikely to see them in Aotearoa New Zealand; and the fruit of *Ficus elastica* is best avoided. The latter is called rubber plant because of its latex sap, but all figs have this feature to different degrees and it is important to be aware that some people can be severely allergic to this sap; so avoid the leaves and do not eat unripe fruit: the figs should be soft, dark and sweet-smelling.

The garden fig is well known so does not need to be described in detail in these pages, and neither do many species that are familiar from the home orchard or supermarket. However, keep an eye out as you may find some that have skipped over into the wild — for instance, the common hazelnut tree, *Corylus avellana*, has naturalised in recent years, so you might be lucky to encounter a wilding bearing nuts.

Another large shrub or small tree commonly grown in the garden that has naturalised in the past decade or so is the camellia. Tea, of course, comes from the fresh-growth leaf tips of *Camellia sinensis*, in particular, though other *Camellia* species are also used. What's more, *Camellia* flowers can be dried and added to various dishes — indeed, the flowers of *Camellia japonica* are a traditional ingredient in Japan. The petals make attractive garnishes, crystallised decorations for cakes or additions to ice cubes.

Both trees and shrubs are often foraged for their berries — and don't overlook gooseberries (*Ribes uva-crispa*), which you might well find wandering out of gardens — but do not assume that all will be edible. The entire clan of the native tutu (*Coriaria* spp.) are poisonous. These species, ranging from small shrubs to small trees, produce dangling stems studded by what look like black berries but are actually tiny flowers wrapped tightly around the seeds. While the juicy petals are edible, the rest of the plant — most importantly, the seeds inside those petals — are lethal. It is a similar case with the flesh of tītoki berries (*Alectryon excelsus*), which is not particularly palatable though deemed edible; however, the seeds contain oils that can hydrolise to hydrogen cyanide. The holly tree, *Ilex aquifolium*, may seem friendly because of its association with the northern hemisphere Christmas, but all parts are toxic, especially those enticing red berries. Various types of privet (*Ligustrum* spp.) pop up everywhere, but the very-dark-purple berries of all varieties are also poisonous, as are the yellow berries on the chinaberry tree, *Melia azedarach*, the black-cherry-like berries of cherry laurel, *Prunus laurocerasus*, the berries of both the endemic ngaio, *Myoporum laetum* and the Australian ngaio, *Myoporum* aff. *insulare*, and the pink berries of the pepper tree, *Schinus mole*. There are other dangerous trees and shrubs, some but not all of which have been touched on in other sections, so be aware that correct identification is essential before you consume your foraging. And, as always, try small quantities at first in case you have allergies.

The warnings can be off-putting, but once you have worked through those elements of identification — shrub or tree, deciduous or evergreen, leaf shape, flowers, berries, aroma — you'll find a whole new array of edible delights out in the wild offering not just figs, nuts, berries, leaves and petals but also sap, bark and pollen. I've not included trees mostly prized for their sap as you need to know what you're doing to tap them without causing damage. Also, maple syrup primarily comes from the sap of the sugar maple, *Acer saccharum*, and a couple of other species (*Acer nigrum*, *Acer rubrum*), none of which have naturalised in this country; you can still get syrup from the various *Acer* species that have spread here, such as box elder (*Acer negundo*), but they have lower sugar content and are not as tasty or productive. However, the seeds of most maple tree species are edible just as they ripen; they have a mild pistachio flavour when cooked (once removed from their wings). The leaves of species such as Japanese maple, *Acer palmatum*, are used as a food wrap in some Asian cultures; and, while not eaten, the leaves of sycamore, *Acer pseudoplatanus*, can be wrapped around certain food as it cooks to impart flavour. The cambium (inner, living bark layer) of species such as *Acer negundo* can be harvested from felled specimens (such as during pest-control work) and dried and ground into a powder to make a sweet flour for baking or as a thickener for a range of dishes.

Interestingly, the silver birch (*Betula pendula*) has a similar range of uses: its sap is likewise edible; and the inner bark from felled trees can also be used in the same way. In addition, birch catkins are edible and contain protein, beta-carotene and trace minerals, but they are are bitter. And live birch twigs, fresh leaves and even pollen can be foraged to make a peppery, minty tea rich in vitamin C.

Though they may look attractive, don't eat these ngaio berries as they're poisonous.

Sweet chestnuts

SCIENTIFIC NAME

Castanea sativa

DESCRIPTION

A medium-sized (5–15m tall) deciduous tree, though in some conditions it can grow much taller. It has a rounded canopy and elongated dark-green leaves with serrated edges. Its edible nuts are encased in a finely spiked husk. Typically each husk has three nuts inside, although often only one will be large enough to be of practical use. The nuts are round or oval in shape, and their shells are glossy brown. The kernels are creamy white, sweet and starchy.

DISTRIBUTION

Throughout Aotearoa New Zealand but more common in eastern South Island.

HABITAT

Areas of moist fertile soils, often in parkland or suburban areas but sometimes on country roadsides too.

SEASONALITY

Mid to late autumn (April–May) for the fully formed nuts.

FORAGING TIPS

Use gloves when picking the husks, which are very spiky. The husks can be rolled on the ground with your foot to get the nuts out. Often only the larger of the three nuts in each husk is of a practical size to remove the thin shell from.

CONFUSION SPECIES

Be aware that horse chestnuts (conkers, *Aesculus* spp.) are poisonous. Their lime-green leaves are divided into distinct lobes growing from a central point, which are very different to the sweet chestnut whose single oval leaves grow up a stem. Also the sweet chestnut has a point on its nut and is inside a husk covered with many fine, spiny needles, which is uncomfortable to hold, while the more rounded horse chestnut has no point on its nut and is inside a fleshy bright-green husk that is visible between its fewer warty spikes and is easier to handle.

Key edible uses

Nuts need to be boiled or roasted to be made edible; the brown skin is astringent and should be removed. They are high in carbohydrates, protein, vitamins and minerals, have a mild, sweet nutty flavour and when cooked are softly textured. Although delicious when consumed as roasted nuts, they are very versatile and can be used in both sweet and savoury dishes, for instance used whole or chopped for stuffing poultry or making nut roasts, puréed to make soups or candies, or ground into flour for baking, thickening or making desserts. The roasted nuts can also be used as an alternative to coffee.

Clockwise from left Chestnut flowers; chestnut forming in its husk; husks on the tree.

Karaka

SCIENTIFIC NAME
Corynocarpus laevigatus

OTHER NAME
Kōpī

DESCRIPTION
A medium-sized to large tree that can reach heights of up to 15m. It has a dense, often rounded crown, a stout trunk and rough, greyish-brown bark. Growing alternately up and around the branches, the dark-green leaves are glossy, leathery, oblong to elliptical in shape, with smooth margins. It produces oval fruits that are 3–5cm long and are initially green but turn yellow then bright orange when ripe. The pulpy flesh surrounds a large seed.

DISTRIBUTION
Primarily found in coastal areas, it is more common in the North Island than the South Island, where it is found mostly in the Nelson, Marlborough and Canterbury regions.

HABITAT
Coastal environments such as coastal bush, dunes and cliffs. Also often used as an ornamental tree planting.

SEASONALITY
Ripe fruits can be harvested from February to April.

CAUTIONARY NOTES
The kernel is deadly poisonous to humans when raw and untreated; but also keep dogs away when foraging as the kernel is toxic to them as well.

Key edible uses

The outer pulp of the ripe fruit — once carefully removed from the kernel — can be eaten raw and has a pleasant, intense, sweet flavour. The orange fruit that has dropped from the trees will be the ripest. Traditionally Māori used the kernels, which had to be heat-treated over several days to remove toxins, to make an edible starch-like flour. See page 498 for details of preparation.

Tree fuchsia

SCIENTIFIC NAME
Fuchsia excorticata

MĀORI NAME
Kōtukutuku

DESCRIPTION
A small deciduous tree typically growing 3–8m tall with flaky orange bark. The leaves are elongated ovals with a dark vein structure. The pendulous purple flowers turn into a fruit shaped like a Tic Tac lolly, which is initially green and then turns dark purple as it ripens.

DISTRIBUTION
Found throughout Aotearoa New Zealand at a wide range of altitudes, though generally not on coastal margins or in drier parts of Central Otago and the East Cape region. Prefers areas with moderate to high rainfall.

HABITAT
Generally found in sunny parts of native forest or bush edges, particularly on hillsides and ridge-lines. It likes sunny aspects and is generally not often found as an understorey species or in wet ground.

SEASONALITY
Ripe fruits are available from December to March.

FORAGING TIPS
A ladder could be useful, as berries are often up high. Also follow kererū as an indication as to where fruiting trees might be. Meanwhile, the imported fuchsias you might have growing in your garden also produce an edible berry, though some species are more tasty than others.

Key edible uses

The edible berry has a distinct floral taste and can be made into jam, added as a garnish to desserts or simply eaten raw.

Walnut

SCIENTIFIC NAME
Juglans regia

OTHER NAME
Persian walnut

DESCRIPTION
A medium-sized to large tree that can reach 30m tall, it has a very wide spread and a rounded crown with sturdy branches and greyish bark. The elongated oval leaves are a vivid green that turns yellow in autumn. Walnuts are enclosed in a hard, thick, wrinkled shell that is surrounded by a smooth, fleshy husk, which is green and freckled but darkens to a blackish colour and splits open when ripe.

DISTRIBUTION
More common in the South Island but grows throughout the country. Most common in Otago, Canterbury, Marlborough, Wellington and Auckland. Largely absent from Southland.

HABITAT
Areas of moist fertile soils, often in parkland or suburban areas, but sometimes on country roadsides and by old settlements.

SEASONALITY
Sap is collected in September to early October. Walnuts can be gathered green in early to mid-December and the ripe walnuts from late February to early April.

FORAGING TIPS
Having a ladder can be useful for accessing fruits higher up on the tree, though they fall when fully ripe so a rake will help you search leaf-litter. There is a dedicated walnut-collecting device available, which has a rotating wire basket on a pole. Use a pin to check unripe green walnuts to ascertain that the shell has not yet formed.

CAUTIONARY NOTES
People with tree-nut allergies should not consume walnuts.

Key edible uses

Nutrient-rich, walnuts are delicious with a dry, sweet, buttery flavour. Often consumed as a raw snack, they are also used in cooking and baking, or puréed to make a nut butter, or chopped to be a pine nut alternative in pesto. Unripe nuts can be foraged when green, just before the shell forms, at which point they can be pickled or used to flavour liquors. The strong aromatic sap from the trunks can be extracted in the spring and used like maple syrup.

Opposite Walnuts grow enclosed in a smooth, green husk. **Inset** A walnut ripening, and breaking out of its husk.

Bay tree

SCIENTIFIC NAME
Laurus nobilis

OTHER NAME
Bay laurel, sweet bay

DESCRIPTION
A small to medium-sized tree, reaching heights of up to 18m in favourable conditions. The dark-green leaves are leathery, lance-shaped, glossy on the upper surface and paler and more dull underneath. When the leaves are crushed they have a savoury aromatic smell. The bay tree produces clusters of small, pale-yellow to greenish-yellow flowers close to the stems. The fruit is a small, olive-like berry that starts green and turns purplish-black when ripe. The berries are not typically consumed but are used for propagation.

DISTRIBUTION
Found mainly around urban centres with limited distribution into the wild. Most common at lower elevations but can be found growing as a shrub in the Queenstown region.

HABITAT
Mainly in abandoned garden environments and especially common in Christchurch's red zone. Likes damp and rich soils in shaded or partially shaded areas.

SEASONALITY
Available year-round.

Key edible uses

Fresh or dried leaves are normally used as a spicy aromatic element in cooking, especially soups and stews, the leaves infusing flavour and then being discarded. The dried fruit can be used in the same way. Dried leaves also make a distinctive herbal tea.

Himalayan honeysuckle

SCIENTIFIC NAME
Leycesteria formosa

OTHER NAME
Pheasant berry

DESCRIPTION
Unlike its distant cousin the climbing honeysuckle, this is a tall, upright shrub, growing to 2m high with a hollow stem having a red tinge. Its leaves are green and oval with a long point. The white tubular flowers grow on long stalks in clusters out of dominant deep-crimson bracts. They produce dark-purple to black berries, which are 4–7mm across.

DISTRIBUTION
Widespread throughout Aotearoa New Zealand although absent from drier parts of Central Otago and alpine zones.

HABITAT
Grows often in thickets as a coloniser species, often on the edge of tracks and roads, shrub patches and native forest as well as in scrubby hill country. Likes areas with moderate to high rainfall and damp soils.

SEASONALITY
Fresh shoots appear in spring, flowers through spring and early summer, with fruits through late summer to early winter.

CAUTIONARY NOTES
Only eat the ripe berries; when unripe they are bitter and potentially toxic.

Key edible uses

Used in traditional Chinese medicine, the berries have a distinct sweet, toffee flavour. The ripe berries when soft are used to make jam and flavour wines; they can be eaten raw or blended into ice cream or juiced and combined with olive oil and honey vinegar to make a zingy salad dressing. The young shoots and leaves can also be eaten.

Magnolia

SCIENTIFIC NAME

Magnolia spp.

DESCRIPTION

There are at least eight different species and numerous cultivars recorded in Aotearoa New Zealand. A medium-sized deciduous tree, though some are shrubby, it has a sprawling structure and elongated oval leaves. The flower buds open in late winter and early spring, in many species before the leaves appear, with a medium to large flower, typically in either a tulip or a star form; the tulip form spreading out its large, fleshy petals. Depending on the variety, the flowers range in colour but are usually white, pink or purple.

DISTRIBUTION

Found throughout Aotearoa New Zealand, generally in parkland and urban environments, in lower-altitude and coastal regions. They tend to be more common from Christchurch north and especially common in the upper half of the North Island.

HABITAT

Parkland and garden environments and occasionally planted as an ornamental roadside tree, but can be foraged in abandoned sections such as Christchurch's red zone and wasteland close to gardens. Limited spread into the wild, but too useful to ignore.

SEASONALITY

Magnolia buds are often fully formed by mid- to late July, the flowers opening from mid-July to early August. Foliage appears over spring and summer and is shed in autumn and winter.

Key edible uses

Magnolia flower petals are edible and have a delicate ginger-citrus flavour. The tulip forms in particular can be used fresh in salads, covered in tempura and fried, as a garnish for desserts, or infused into teas, rice dishes or syrups, vinegar and oils to add a unique floral essence. The flower buds can be pickled to make a tangy caper. In Japan the leaves of a certain species of magnolia are also consumed, but it is not a variety found here.

Kawakawa

SCIENTIFIC NAME
Piper excelsum

OTHER NAME
Macropiper excelsum

DESCRIPTION
This shrub or small tree typically grows to 5m high. Its heart-shaped, bright-green leaves grow up to 10cm in size on spindly branches; they have definite veins and are often riddled with holes created by the kawakawa looper caterpillar, *Cleora scriptaria*. The small flowers are indistinct but produce striking fruit, 1–4cm long, which are green upright fingers that turn a vivid orange as they ripen.

DISTRIBUTION
Found throughout the North Island and upper half of the South Island. Generally only found in coastal areas in the South Island but more widespread further north.

HABITAT
Grows as a subcanopy shrub underneath larger trees, generally in native forests but also in pine forests. Prefers sheltered and shaded frost-free environments.

SEASONALITY
Leaves available year-round and fruits from December to May.

Key edible uses

Used in traditional Māori medicine, researchers have confirmed that this plant contains over 60 biologically active compounds. The leaves have a very refreshing peppery and slightly numbing sensation on the palate. They can be eaten raw but have a strong flavour, so are best steamed or pickled or used in a tea. The dried leaves and green fruit can be blended with sugar or with salt to make sweet or savoury seasonings. The flesh of the orange fruit is sweet, while its seeds are intensely aromatic. The ripe fruit can be eaten raw, and when added to jams brings a touch of spice and heat.

Inset Kawakawa fruit is bright yellow and finger-shaped. **Opposite** Kawakawa leaves are used in Māori medicine, and make delicious tea. When picking, choose the leaves with holes in them, as the kawakawa looper caterpillar releases active compounds when munching!

Lemonwood

SCIENTIFIC NAME
Pittosporum eugenioides

MĀORI NAME
Tarata

DESCRIPTION
A moderate-sized tree that can reach 12m tall and often has a sticky gum exuding from its trunk. Its elongated, bright-green leaves have a paler mid-rib and wavy edges. The fragrant flowers are yellow and borne in clusters that produce bunches of green berries that turn yellow and split to reveal hard black seeds.

DISTRIBUTION
Found throughout the country at a wide range of altitudes, it is one of our most widespread trees, but is rare along the southern half of the South Island's west coast.

HABITAT
Often found on the edge of forests. Also very common as a planted species in urban environments.

SEASONALITY
Available year-round. Flowers in spring with September being a peak month.

FORAGING TIPS
Mature leaves have the best flavour and can easily be handpicked. As an added bonus, velvet shank mushrooms (see page 458) are often found on dead and dying lemonwood trees.

Key edible uses

The citrus-flavoured leaves are not pleasant eaten raw but can be used in a similar way to a bay tree leaf, added to cooking soups and stews and discarded. Both leaves and flowers can be used to infuse a lemon taste into syrup, vinegar and alcohol.

Inset Lemonwood flowers are fragrant and add a lemony taste to syrups and alcohols. **Opposite** Lemonwood leaves don't make great eating when raw, but can be added to soups and stews as you would a bay leaf.

Horopito

SCIENTIFIC NAME

Pseudowintera colorata

OTHER NAMES

Mountain horopito, pepper tree, puhikawa, ōramarama, ramarama

DESCRIPTION

Growing up to 5m high, this tall shrub or small tree has grey branches with white blotches. Depending on the amount of light available, the wavy oval leaves are generally green but might have red or purple margins, mottling or spots, looking a bit like rust. The small flowers are indistinct and produce black berries that are small and round.

DISTRIBUTION

Found generally inland above 300m altitude unless in the lower South Island or in garden plantings at lower altitudes, and throughout Aotearoa New Zealand except north of Auckland.

HABITAT

Likes open environments and well-drained soils and is often found on native forest margins, on sloping ground.

SEASONALITY

Foliage available year-round, berries appearing from January to April with a peak in March.

FORAGING TIPS

The leaves vary: leaves in shaded locations tend to be bright green whereas those in bright sunlight go deep crimson and have a very intense flavour. Leaves grow densely on the outer branches and can be stripped off by rubbing your hand along the branch. It is advisable to wear a gardening glove when doing this. A small hand-saw can be used for cutting branches, but collect the sawdust as it can be used to smoke foods.

CAUTIONARY NOTES

Very hot, especially the berries, so use carefully.

Key edible uses

This indigenous pepper has medicinal qualities and a hot aromatic flavour, which is mainly used as a flavouring for both savoury and sweet dishes. The spicy oils in the leaves are released when used like a bay leaf and added to stocks, soups and sauces or dried and ground to make a spice rub. The leaves are excellent for infusing flavour into oils, vinegars and spirits such as gin. The wood and dried leaves can be used to smoke different foods and when mixed with mānuka (see page 184) make a dynamic smoke flavour. The ripe black berries are very hot and can be used in infusions and sauces, but are at a 10/10 on the heat scale so use with care.

Clockwise from left Ripe horopito berries; sun exposed leaves develop a bright-red speckling; mature shrub; red leaves.

Oak

SCIENTIFIC NAME
Quercus spp.

DESCRIPTION
There are about 50 species of *Quercus*, though only a handful are present in Aotearoa New Zealand. They are large trees growing 6–25m tall. While most are deciduous, some species are evergreens. Known for their sturdy trunks, strong branches and dense, spreading canopies, their leaves are typically lobed. The tree produces an elongated or rounded acorn nut.

DISTRIBUTION
Throughout Aotearoa New Zealand but more common in eastern South Island.

HABITAT
Areas of moist fertile soils, often in parkland or suburban areas but sometimes on country roadsides and established farms and homesteads.

SEASONALITY
Ripe acorns can be gathered from late summer and through autumn. Fresh growth is gathered in spring.

FORAGING TIPS
A rake can be useful for gathering acorns.

CAUTIONARY NOTES
While some people might consume the young oak leaves, they contain tannins so are best avoided. Likewise the acorns must be processed properly to make them edible.

Key edible uses

Acorns can be processed to make flour, which has to be done thoroughly to remove the tannins (see page 498). The resulting acorn meal can be used as a flour substitute for baking or as a coffee substitute. You can make oak-leaf wine, though more commonly the leaves are used for herbal tea, sometimes incorporating fresh-growth twigs, which provide an earthy flavour; the leaves contain potentially high-levels of tannins, but the tea would taste revolting at dangerous levels. Oak wood and bark can be used for smoking foods; after all, it is used to make barrels for ageing alcohol.

Staghorn sumac

SCIENTIFIC NAME
Rhus typhina

OTHER NAME
Velvet sumac

DESCRIPTION
A small to medium-sized deciduous shrub or small tree with fine, upright, sprawling branches. Its large leaves are made up of separate leaflets that are elongated spears with serrated edges and run in pairs up the mid-rib. They have a citrus aroma when crushed, and are dull green turning a spectacular red in autumn before being shed. The stems and stalks are covered in red hairs, like the velvet on a stag's antler. It has fuzzy upright 'flames' made up of red berry-like fruit, technically drupes. The fruit will stay on the tree into winter, its colour becoming progressively duller.

DISTRIBUTION
Grows throughout Aotearoa New Zealand, mainly near settlements. Absent from high-rainfall zones on the West Coast and in alpine regions.

HABITAT
Mainly found near areas of habitation but also grows in abandoned green spaces. Will tolerate a range of soil types, is quite frost-tolerant and occurs in both inland and coastal locations.

SEASONALITY
The ripe fruit/drupes are available from late January to June.

FORAGING TIPS
The fruit can break apart quite easily, so use a sharp knife to harvest them.

CAUTIONARY NOTES/CONFUSION SPECIES
Only harvest when the fruit is fully ripe and red all the way through. The fruit is high in tannins, so use in moderation.

The fruit stems contain a milky substance that may cause skin allergies for some people, though these concerns are possibly from a confusion with *Toxicodendron succedaneum* (also known as *Rhus succedanea*), which has very similar leaves and similar upright fruit but produces light-brown berries rather than red; its sap is well known for provoking allergic reactions. The poisonous species of sumac is *Toxicodendron vernix*, which has white fruit under its fuzzy coating, but it is not known in this country.

Key edible uses

The fruit is high in vitamins A and C, has a citric flavour and can be fermented to make an alcoholic beverage, used to flavour kombucha or merely infused in water for half an hour to make a tangy cordial or jelly, sweetening to taste. Do not use hot water with the fruit as this will bring out the tannins and accentuate the acidity. The fruit can also be dried or dehydrated, then ground and sieved to remove the acidic seeds, and used as a spicy seasoning for a range of Moroccan-style dishes.

Blackcurrant

SCIENTIFIC NAME

Ribes nigrum

DESCRIPTION

Growing a little over 1m high, it is a scrubby bush and can form thickets. The dark-green leaves have three to five lobes and finely serrated margins. The leaves are aromatic when crushed. The flowers form small yellowish to light-brown clusters, producing the dark-purple blackcurrant berries, which are typically 1cm in diameter. They have a smooth glossy skin and contain numerous tiny seeds.

DISTRIBUTION

A patchy distribution within Southland, south-east Otago, the Catlins, Central Otago, central Canterbury and parts of Taranaki. Primarily found in the southern half of the South Island.

HABITAT

Found on the edge of settlements and abandoned homesteads. Grows among matagouri shrub (*Discaria toumatou*) at some sites in Central Otago and often next to lakes or rivers. Likes frost-free areas with good sunlight, but only thrives with cool air temperatures over summer.

SEASONALITY

Fresh-growth leaves are gathered in spring, with fruit ripe from December to mid-January.

FORAGING TIPS

Take a small container to collect your berries to prevent them being crushed. The berries grow in bunches and can be raked off by moving your hand along the stem.

Key edible uses

High in vitamin C, the ripe berries are juicy and have both a tart flavour and a subtle sweetness. The ripe fruit can be eaten raw or cooked and added to both savoury and sweet dishes, including pies, jams and jellies. Leaves are best foraged in spring and make a good tea or addition to soups and stews. In springtime the flowers will provide a flavour infusion (although the flowering currant, see page 401, is better for this application).

Flowering currant

SCIENTIFIC NAME

Ribes sanguineum

DESCRIPTION

A medium-sized shrub that is taller than the blackcurrant (see page 400), reaching up to 3m high with a similar spread. It has multiple stems arising from the base, and a rounded form. The leaves are similar to the blackcurrant, with three to five medium- to dark-green lobes and serrated edges, arranged alternately along the stems. The most striking feature of flowering currants is the abundant and showy drooping clusters of bright-pink flowers. They are tubular with five petals around a central circle of white stamens.

DISTRIBUTION

Found throughout Aotearoa New Zealand often on the edges of foothill scrub or near cities and towns where the species has established as a garden escapee.

HABITAT

Likes sunny areas with well-drained soils.

SEASONALITY

Leaves are best harvested in early spring with flowers harvested later in the spring.

Key edible uses

The aromatic, fruity flowers are strongly flavoured and are ideal for infusing into syrups, jellies, cordials, spirits, vinegars and oils. The leaves can be dried to make herbal tea. The small fruit can be consumed but are often tasteless, but being high in pectin they are useful additions when making jam, jellies or chutney.

Elderflower

SCIENTIFIC NAME
Sambucus nigra

OTHER NAMES
Elder, elderberry

DESCRIPTION
A small to medium-sized tall shrub or shrubby deciduous tree, it produces plate-like clusters of tiny fragrant white flowers that turn into clusters of small round green berries that progressively darken to purplish-black. The leaves are made up of two to three pairs of serrated, oval, pointed leaflets with a single leaflet on the end.

DISTRIBUTION
Found throughout central and eastern South Island and the Wellington region, with only patchy distribution further north.

HABITAT
Often in hedgerows along roadside margins in regions with low to medium rainfall. Tends to prefer cooler conditions with strong seasonal variation.

SEASONALITY
The flowers are harvested from late September until mid-December and the fruit from mid-February until the end of April.

FORAGING TIPS
Choose a spot well away from traffic fumes. Choose the fully opened elderflowers, shaking out any insects.

CAUTIONARY NOTES
It is important to remove the leaves and green stems, which are mildly toxic unless boiled. Elderberries contain cyanogenic glucosides, so must be cooked to remove the toxicity. Do not eat raw.

Key edible uses

The flowers have a sweet aromatic flavour that is popular for making wines, syrups and cordials. The flowers can also be added to a range of desserts, stewing fruit and jams, and can be dipped in light batter and fried until crisp. They can even be dried for a tea. The ripe berries have an intense berry flavour that is quite acid so is best used in moderation and paired with sweet ingredients to make wine, jams, jellies, chutneys and pies. The berries can be dried as long as they are cooked when used. Green elderberries can be pickled and made into capers.

Clockwise from left Ripe elderberries; unripe berries; elderflower shrub; mature elderflowers.

Seaweeds

Wakame

405

Aotearoa New Zealand has over 900 species of seaweed, of which 25 are of practical use to foragers. The majority are red seaweeds (*Rhodophyta*), followed by brown (*Phaeophyceae*) and green (*Chlorophyta*). Overall, seaweeds are a safe group of plants to forage as only a few species contain harmful toxins. Most of the recognised edible seaweeds in Aotearoa New Zealand are brown. The culinary use of many of the red seaweeds is relatively unknown, so there is a large amount of room to explore how best to prepare them.

Seaweed usually consists of a holdfast (the root-like organ attaching the plant to a surface) and the stipe (the equivalent of a stem or trunk), which might be bare or be part of the blade (the leaf of the seaweed). Some brown seaweeds also have air bladders, a bead-like float.

Rocky coasts and shallow reefs are the main areas for seaweed foraging, but a few species also occur in estuarine locations. Picking at extreme low tides or undertaking shallow-water snorkelling will increase the diversity of seaweed species you can forage. It is important to forage seaweeds from areas with clean water and avoid foraging after periods of heavy rainfall. Also, the harvesting of seaweeds is controlled by the Fisheries Act and care needs to be taken to check any local regulations, especially within mahinga mātaitai or taiāpure areas. But overall, recreational wild harvesting of seaweeds is allowed along most of our coastline. Ideally, seaweeds should be harvested while they are still growing, but seaweeds freshly washed up on the beach can also be gathered.

Once harvested, most seaweeds can be dried to extend their shelf-life and improve their flavour. Seaweeds are popular for foraging, being a healthy addition to our diet and many having an umami flavour profile. However, because of restrictions on commercial supply, they have been under-represented in our cuisine. They can be used in many ways, such as sprinkles, sushi, food wraps and added to soups, stews and salads.

Overall, six types of seaweed make up the bulk of edible seaweeds in Aotearoa: bull kelp, bladder kelp, wakame, karengo, sea lettuce and New Zealand kombu, *Ecklonia*. There is only one mildly toxic seaweed that should be avoided because of its high acid content: flattened acid kelp, *Desmarestia ligulate*, a brown seaweed with a bumpy, fleshy holdfast, a main rib and leafy-looking lobes growing off its length, which can look fern-like. However, species new to Aotearoa can easily spread and take hold, so be wary of trying anything you cannot clearly identify.

Seaweeds contain many nutritious elements, including a number of vitamins and minerals, particularly iodine, which plays an important part in thyroid health (as long as it is not consumed in excess). Seaweeds are rich in other compounds that aid gut health and, although research is still in its infancy, there are indications that many also have significant antibacterial and antioxidant properties.

Green seaweeds are generally high in protein, mineral, iron, vitamins and fibre. Red seaweeds tend to be higher in protein than other kinds of seaweeds are a wonderful source of essential amino acids, iron, iodine and fibre.

There is great diversity among red seaweeds, and many can be eaten, especially the softer-textured ones. Our red seaweed species are under-utilised because many of them occur in deeper water environments. They're mostly available to foragers only when freshly beach-cast or growing on ropes of mussel farms, or to those who forage by scuba diving. Keep an eye out on the beach and be prepared to experiment. Rarer in our waters but worth finding is the wonderfully named Turkish towel, *Chondracanthus exasperatus*, which is not generally deemed edible but is used as a gelling agent — and even as a mineral-rich flannel.

Brown seaweeds are also high in iodine and contain omega-3 and omega-6 fats and important trace elements. Many brown seaweeds, such as common flapjack, *Carpophyllum maschalocarpum*, have a tough texture and a bitter taste when fresh, and although edible are rarely used in Aotearoa New Zealand. For some species, such as those in the *Marginariella* genus, the bladders have a good flavour when both fresh and pickled, though the rest of the plant is tough. Many of the other brown seaweeds can be boiled down to make a dashi stock or used to wrap food to provide a seasoning. There is a lot of scope for experimentation.

Key seaweed foraging tips

- Pick during low spring tides, especially in areas with turbid waters, when the greatest diversity of seaweeds will be exposed.
- Be mindful of the water quality of the site you forage from: avoid foraging seaweeds after heavy rainfall; avoid sewage outlets; avoid areas with high levels of human activity; avoid areas close to intensive agriculture; and avoid areas where heavy metals may have been discharged, for instance by nearby industry, as they may be absorbed by the seaweeds.
- Avoid seaweed that has algae growing on it.
- Seaweeds freshly washed onto the beach are safe to use as long as they do not smell or have blotchy discolouration.
- Use a catch bag that freely drains to collect seaweeds.
- Use a sharp knife to pick off seaweed tips.
- Wash thoroughly in clean saltwater to remove grit and any snails. A salad spinner can be useful.
- Consider snorkelling and having a wetsuit, especially during the cooler months.
- Keep seaweeds chilled after harvest as they can decompose quickly.
- Dehydrate the harvest to last through the year.

Young bull kelp

Red straw weed

SCIENTIFIC NAME
Agarophyton chilense, formerly known as *Gracilaria chilensis*

MĀORI NAME
Kapoke whero

DESCRIPTION
Fine hair-like stems that are cylindrical filaments. True to red seaweed's tendency to variety, you cannot rely on colour to identify this species as it might be purple-red or yellow-green or brown. Likewise, it might be a short thicket of fine branches or sparse and stringy, growing to 20cm.

DISTRIBUTION
Throughout Aotearoa New Zealand.

HABITAT
Mainly found in large estuaries and shallow harbours with soft sediments.

SEASONALITY
Available year-round.

FORAGING TIPS
Collect at low tide. Has a good shelf-life and will keep fresh in the fridge for a few days.

CAUTIONARY NOTES
Carefully remove any stones or shells the plant is attached to, and wash off all fine sediment.

Key edible uses

Add to a salad, either raw or boiled, or dehydrate to use as a vegan gelatine. Its thickening properties are also useful for soups — and in soap-making.

Inset Dehydrated red straw weed with its hair-like appearance.

Sea rimu

SCIENTIFIC NAME
Caulerpa brownii

DESCRIPTION
Grows as small leaves up to 20cm long, with fine spiky blades coming off a central stem. Dark-green colour with a size and structure very similar to the foliage of a rimu tree.

DISTRIBUTION
Throughout Aotearoa New Zealand but with a patchy distribution along open coastlines.

HABITAT
Grows in areas of reef of moderate wave exposure. Found below the intertidal zone to 10m depth.

SEASONALITY
Available year-round.

FORAGING TIPS
Tends to grow in clusters in gutters in reefs, so pick a calm day to harvest, though it can be gathered after storms when freshly washed onto the beach. As it grows below the intertidal zone, snorkelling gear is useful when harvesting. Having a fine-meshed catch bag to place the seaweed in is advisable too.

Harvest with care in small amounts as this species naturally only occurs in limited quantities.

Key edible uses

Eaten fresh, it has a soft texture and distinct flavour and is good added to a salad. Can be dehydrated and made into a powder too.

Sea grapes

SCIENTIFIC NAME
Caulerpa racemosa

DESCRIPTION
Multiple small (about 3mm) olive-green beads in tight clusters coming off a central stipe.

DISTRIBUTION
Found from Marlborough Sounds north to Cape Rēinga. Very localised patches due to specialised habitat requirements. Grows in areas of reef of moderate wave exposure. Found below the intertidal zone to 5m depth. Often growing in rocky gutters.

SEASONALITY
Available year-round but is in optimal condition over the winter months.

FORAGING TIPS
The same tips as for sea rimu (see page 409). Over the hot summer the tips can fade and are best not harvested.

CAUTIONARY NOTES
Take extreme care when foraging in exposed rocky gutters. Wearing a wetsuit is advised.

Key edible uses

Described as vegetarian caviar with a unique flavour, it is tasty to eat fresh or preserved in a light pickle or vinegar. A good addition to salads. Being highly perishable it needs to be eaten or processed soon after harvest.

Sea emeralds

SCIENTIFIC NAME
Chaetomorpha coliformis

DESCRIPTION
Grows as a string of bright-green beads (3–4mm diameter) in a single linear strand.

DISTRIBUTION
Throughout Aotearoa New Zealand, but especially common on the Otago coastline.

HABITAT
Below the intertidal zone so needs to be dived for, often in depths from 1–7m. Commonly grows as an epiphyte on the base of sea tulips with the strands in clusters.

SEASONALITY
Available year-round.

FORAGING TIPS
Best snorkelled for. Easily damaged, so best put straight into a container in a catch bag.

Key edible uses

Has a pleasant salty, cucumber-like flavour. Eat raw as a garnish or in salads.

Branching velvetweed

SCIENTIFIC NAME
Codium fragile

OTHER NAMES
Dead man's fingers, green sponge, felty fingers, forked felt-alga, sea fingers, sponge seaweed, stag seaweed

DESCRIPTION
A dark-green seaweed with a velvet texture. The multi-branched fronds grow up to 30cm long and 1cm wide from a spongy holdfast.

DISTRIBUTION
Throughout Aotearoa New Zealand coastal waters.

HABITAT
Mainly in sheltered areas of coastline, such as harbours, embayments and estuaries. Needs stones or rocks to anchor on but can occasionally also be found in open sandy-bottomed areas attached to dead shells. Found from the lower intertidal zone to about 5m depth.

SEASONALITY
Throughout the year, though can die back in autumn.

FORAGING TIPS
Mainly gathered at low tide, growing on shells or rocks. Cut the outer fingers off, leaving some of the plant attached to regrow. Can often be gathered when freshly washed onto the beach after stormy weather.

Key edible uses

Can be eaten raw, chopped up and blended into a salad. It is also very tasty when dehydrated and ground into a powder to be used as a seasoning with a strong umami-oyster flavour.

Bull kelp

SCIENTIFIC NAME
Durvillaea spp.

MĀORI NAME
Rimurapa

DESCRIPTION
There are various species — such as *Durvillaea antarctica*, *Durvillaea chathamensis*, *Durvillaea poha* and *Durvillaea willana* — all with similar characteristics: large, thick and broad blades, which have a honeycomb texture and are well adapted to areas of surging waves. A thick fleshy stem with a large divided blackish-brown blade and a broad holdfast are key features. Several of these species are endemic to Aotearoa New Zealand. Note that *Durvillaea antarctica* and *Durvillaea poha* have been declining and have recently been viewed as at risk, so beach-harvest bull kelp only.

DISTRIBUTION
Found throughout Aotearoa New Zealand but most common in Otago and Southland.

HABITAT
Open, wave-exposed coastlines with rock platforms and reefs to anchor on.

SEASONALITY
Available year-round.

FORAGING TIPS
Fresh plants can be beach-harvested after stormy weather. Rinse in cold water as heat will release its flavour.

CAUTIONARY NOTES
Bull kelp is regarded as a taonga by Māori, so respect local protocols. Crustaceans and other creatures like to live on and around the holdfast, so clean them off when collecting. Bull kelps can be infected by parasites: check for galls or other signs of infection and only harvest healthy specimens. Cook thoroughly.

Key edible uses

While its tough texture is challenging, when processed, bull kelp has a wide range of culinary uses and is particularly high in iodine. The base, blades and stem can be boiled down to make seaweed stock with a robust, beefy flavour — a great option for getting richness into vegetarian dishes. The younger, narrow blades are best to cut into chips and roast in the oven. The fleshier parts make chunky seaweed chips when fried. The fully mature large blades can be opened up to make an oven bag; food cooked inside will be infused with a rich umami flavour.

Opposite Bull kelp is a taonga for Māori, so respect any local rules before harvesting.

New Zealand kombu

SCIENTIFIC NAME
Ecklonia radiata

OTHER NAMES
Leather kelp, paddleweed, spiny kelp

DESCRIPTION
Golden-brown plants grow to 1m high and consist of a holdfast made up of cylindrical fingers that attach to surfaces, while the stipe, which is also cylindrical, is unbranched. It has a single large blade, divided into lobes. Blades may be smooth or wrinkled.

DISTRIBUTION
Throughout Aotearoa New Zealand coastal waters, but more common and a dominant brown seaweed in the North Island. Tends to be in slightly deeper water in the South Island.

HABITAT
Reefs and rocky areas with some wave exposure, mainly along open coastlines at a depth of 2–20m.

SEASONALITY
Available year-round.

FORAGING TIPS
Often washed onto the beach in large amounts and can be harvested when fresh. Take the time to wash any sand or stones off the fronds.

Key edible uses

Blades can be dehydrated and made into a sprinkle. The stems can be pickled in vinegar. When cooked, the blades and leaves can be finely shredded and put into a seaweed salad. When dried, kombu makes an excellent broth.

Inset Mature kombu has a flattened centre, and yellow-brown colour. **Above opposite** Kombu can be foraged when freshly beach washed. **Below** Forage after a storm, as large amounts of kombu can show up on the beach.

Neptune's necklace

SCIENTIFIC NAME
Hormosira banksii

DESCRIPTION
A distinctive bead-shaped seaweed.

DISTRIBUTION
Throughout Aotearoa New Zealand wherever there are flat rock platforms. Tends to be far more common on eastern coasts.

HABITAT
On rocky platforms in the mid to lower intertidal zone. Often found outside the normal range of beach-cast.

SEASONALITY
Available year-round.

FORAGING TIPS
Best to cut off the outer newer beads. Avoid removing whole plants, to allow regrowth.

Inset Fresh Neptune's necklace has the best flavour. **Above opposite** Fully grown Neptune's necklace. **Below** This seaweed prefers to grow on flat rock forms, growing in dense beds.

Key edible uses

The beads can be eaten and have a mild flavour. Slice or pierce the beads prior to cooking, otherwise they explode. They can be lightly blanched and will turn bright green. The blanched beads are suited for pickling in a light vinegar to add to salads. The whole plant can be eaten, but choose young plants. Best harvested in late winter and early spring when the seaweed is most tender. The new growth can be included in a salad, adding a flavoursome crunchy texture. It can also be pickled in white vinegar. High in iodine.

Bladder kelp

SCIENTIFIC NAME
Macrocystis pyrifera

OTHER NAME
Giant kelp

DESCRIPTION
Grows prolifically and quickly to many metres in length. A brown seaweed with thin stipes and large, flat, thin and elongated oval-shaped blades, with a single distinct oval-shaped bladder at the bottom of each blade.

DISTRIBUTION
From Wellington south to Stewart Island. Absent from most of the North Island. Most common on the South Island's east coast.

HABITAT
Likes areas that are sheltered from strong wave surges but also with a good current flow, so generally found in the mid and outer sections of bays and harbours or growing behind rock breakwaters.

SEASONALITY
Available year-round, but often in best condition in spring.

FORAGING TIPS
A sharp knife is required to cut off at the base. Fresh plants can be beach-harvested after stormy weather.

Key edible uses

Bladder kelp, with its rich salty umami flavour, is one of my favourite edible seaweeds. Harvesting the fresh leaves near the tip of the stem is recommended, as these provide delicious eating. When fresh, the leaves can be cut into strips and slightly blanched, turning them bright green, to be added to a salad. Alternatively, the blade-shaped leaves can be dried and used to make broth or ground down to make a seaweed pepper — which is also commercially available. Rapid drying is best done over summer when a warm north-westerly wind allows speedy dehydration. The pepper will stay fresh for a year when stored in an airtight container. It can be mixed with sea salt in a salt grinder to add a distinctive, slightly peppery taste that is ideally suited to many seafood dishes. If dried and kept intact, the leaves can be crisped in the oven for a short period of time with a little olive oil for extra taste and served as seaweed chips. Note that many types of brown seaweed will turn a bright-green colour when lightly blanched, adding to their appeal in both taste and edibility.

Above opposite Bladder kelp grows to many metres in length and has a distinct oval shaped bladder. **Inset** The bladders are suspended at the end of the broad blades of seaweed. **Below** Rapidly air drying this kelp provides the best results.

Karengo

SCIENTIFIC NAMES
Pyropia spp., *Pyropia columbina*

OTHER NAMES
Nori, parengo, southern laver

DESCRIPTION
Around 35 karengo species have been identified in Aotearoa New Zealand to date, including the most recognised one: *Porphyra columbina*. Uniquely, the species is only one cell thick. They come in many shapes and textures, from tough to tender, some looking like thin plastic wrappers that have been melted, others like long wavy ribbons. Their colour is equally varied, and can also have gold and green hues. Karengo is often found naturally dried out on the upper part of the intertidal shore.

DISTRIBUTION
Widespread on rocky and stony shorelines, from sheltered to moderately exposed coastlines, such as Kaikōura, to harbour environments.

HABITAT
Occurs in the intertidal zone and can often be found in mid-intertidal areas, naturally dried but still alive, clinging tightly to rocks, especially sedimentary ones. Karengo occurs in dense mats on rock surfaces when in peak season.

SEASONALITY
The main season is from June to November, but small amounts can be foraged year-round.

FORAGING TIPS
Use a sharp knife to cut the blades of karengo, leaving a small amount of growth on the base so that the blade can regrow (even if it looks dried up and dead, it is likely to be alive).

Key edible uses

The texture can be tough when fresh, so it is best dehydrated and crisped to improve edibility. It has a very strong umami flavour; a similar species is used internationally for the production of sushi sheets. Flavours develop well when lightly roasted, making it a delicious snack. You can also cook it in butter and spread it on bread, steam it in a hāngī, toast it and crumble it into soups or stocks, or turn it into a seasoning by drying, grinding and mixing it with salt. There is an Irish recipe for carrageen moss pudding which combines the moss with milk, sugar and an egg: dried karengo is a good substitute.

Deadman's fingers

SCIENTIFIC NAME
Splachnidium rugosum

OTHER NAME
Brown gummyweed

DESCRIPTION
A small tubular seaweed that grows in clusters up to 8cm long. Golden brown in colour, it has a clear gelatinous fluid inside.

DISTRIBUTION
Widespread throughout Aotearoa New Zealand but more common on the eastern coasts.

HABITAT
Rock platforms of moderate exposure and often found along open coasts, but largely absent from those with high wave exposure.

SEASONALITY
Available year-round.

Opposite inset Karengo is often an olive-green colour and has a cellophane-like appearance. **Above inset** Its brown, tubular form makes this species very distinct.

Key edible uses

Despite its unappetising name it can be eaten raw, making the most of its interesting gelatinous texture. When lightly blanched, it provides a point of interest as a garnish on seafood dishes.

Wakame

SCIENTIFIC NAME
Undaria pinnatifida

DESCRIPTION
Grows as a large single blade or leaf with a central stipe. It has a blackish-brown colour, and its holdfast at the base has a distinctive frill-like structure where the seaweed will anchor onto rocks, wharf pilings, boat hulls or ropes.

DISTRIBUTION
Found throughout Aotearoa New Zealand, but has a patchy distribution that correlates with areas where boats have moored as this introduced seaweed has a poor natural mechanism of dispersal. It has been declared an invasive species in Aotearoa New Zealand so harvesting it is doing the environment a favour.

HABITAT
Often found in harbours, bays, mussel rafts and areas with sheltered coastline. Requires either rocks or artificial structures to anchor on.

SEASONALITY
An annual species with new growth in autumn; best harvested in winter and early spring. Dies away over the summer.

FORAGING TIPS
A sharp knife is required to cut off at the base. Once collected, use quickly when fresh (as wakame has a short shelf-life).

Key edible uses

One of our most versatile and useful seaweeds. Has a soft texture and a rich savoury flavour. It is often blanched and made into a seaweed salad. It is also very good when crisped up and dehydrated, and often used in miso soups. Can be roasted and used as a seaweed wrap. The frill part is very tasty when roasted in the oven. A top seaweed for drying and making into a seaweed sprinkle, and also good to use in ferments. It is particularly high in omega-3 fats and various minerals.

Above opposite The base of wakame has a rubbery frilly structure. **Below** The plant grows as a dark brown frond with a distinct central stipe.

Broadleaf sea lettuce

SCIENTIFIC NAME

Ulva lactuca

DESCRIPTION

A bright-green seaweed that grows in large flat but ruffled fronds. Has a thin leaf with a cellophane-like appearance.

DISTRIBUTION

Found throughout Aotearoa New Zealand but with a very patchy distribution. Where it occurs, it can be very abundant.

HABITAT

Often in estuaries, harbours, bays, mussel rafts and areas with sheltered coastlines. Requires rocks, shells or artificial structures to anchor on.

SEASONALITY

Available year-round but often in best condition in winter. However, it can also grow in large amounts over summer in estuaries.

FORAGING TIPS

Use a sharp knife to cut near the base off rocks. It is best rinsed in saltwater to maintain the flavour. Make sure that the location has clean water. Once gathered, it can be frozen or dehydrated for later use as it has a good shelf-life. Colour fades easily so best stored in a dry, dark location.

CAUTIONARY NOTES

In summer, rotting sea lettuce can contaminate the harvest.

Key edible uses

A very popular seaweed globally with a very distinct oyster-like flavour, which is enhanced when the seaweed is dehydrated or lightly roasted. It tends to be quite tough when fresh, but steaming or quick frying will soften the texture. As it grows in a thin sheet, it is also useful as a food wrap for cooking seafood in. Can be lightly blanched and shredded, served with a light oil and sesame seeds. Can also be put into fritters, or dried, ground and sprinkled into stews and soups.

Inset Fresh growth sea lettuce is unique among seaweeds for its green colour, and form. **Clockwise from left** A fully mature sea lettuce plant; freshly foraged growth; a productive rocky coast, with sea lettuce growing on rocks.

Fungi

A diversity of edible fungi

In Aotearoa New Zealand less than 1% of the species we have are considered edible. In total there are around 120 edible species of fungi here, although a selection of 25 species make up, at a practical level, about 95% of the fungi foraged. I will focus on those species in this book. Māori recognised about 12 types of edible fungi; the majority of fungi now foraged in this country are introduced species.

Fungi are a rich source of protein, essential amino acids, vitamins, minerals, fibre and essential polysaccharides. There is no doubt that their nutritious value has made them increasingly popular for foraging along with their ephemeral nature, ornate forms and often distinct and sought-after flavours.

Fungi Forms

Fungi can be divided into four main groups: gilled, bolete, bracket, and the remainder that come in myriad forms, such as coral fungus, basket fungus and puffballs, and oddities like beech strawberries (*Cyttaria gunnii*) and lichens.

Boletes tend to have pores on the underside of the cap rather than gills and grow symbiotically with trees (a relationship known as mycorrhizal). Most edible boletes in New Zealand are introduced species; native boletes tend to have bitter and unpleasant flavours; and many are specific to the types of tree they grow next to. Birch boletes, for example, only grow under birch trees, so finding the correct tree is important for locating these mushrooms. The most popular and flavourful of the boletes, the porcini, associate with a wide range of introduced trees.

There are at least 25 species in the gilled mushroom genus *Agaricus*. The most well-known are the common field mushroom and horse mushroom. The salty mushroom (*Agaricus bernardii*) grows in low-lying areas with slightly saline soils in the South Island. Similar in appearance and gill colour to field and horse mushrooms, the cap becomes rough and scaly as it matures. It has a distinct purplish colour when bruised or cooked, and is large like a horse mushroom.There are many species of *Agaricus* with brown scaly caps and dark gills that are edible, and some species occur near trees rather than in open fields. These are known as 'brown scale cap' mushrooms and have the aromatic *Agaricus* mushroom smell, like the blushing wood mushroom (*Agaricus sylvaticus*). There are many lesser-known *Agaricus* species that are still awaiting full scientific description, and many do not have common English names — a full list of these species can be found on iNaturalist (see page 513).

Bracket fungi, also known as shelf fungi, grow on living trees, hollowing them out, and are often found on trees or stumps that have little life left in them. They don't have gills but distribute their spores through pores and are generally either spongey or hard. It's best to forage the younger specimens.

Puffballs grow as stalkless balls, and when mature, release their spores in a puff. The giant puffball is the forager's prize, but you can also eat the common puffball (*Lycoperdon perlatum*) which is smaller. Its surface is covered in tiny, raised, pyramid-shaped spikes. Be aware that some toxic fungi also form balls, such as earthballs. Young gilled mushrooms can also start out as balls.

There are hundreds of species of mushrooms in all sorts of forms and colours. Yeast, lichen and moulds are also part of the wider fungi family, so the species covered here are merely a brief introduction to these fascinating organisms.

Where to find fungi

Fungi are commonly found in areas with warm moist microclimates, often with clustered distributions. When you find one species, others are likely to be nearby. Some toxic but visible species, such as the fly agaric, are markers that the environmental conditions are good for fungi in general. Fungi tend to re-occur in the same location year after year, so marking locations using GPS is useful — especially in large, forested areas.

Avoid leaving signs in the field (such as shavings from a mushroom stem when the dirt was removed), as fungi foraging is competitive and you want to minimise any signs that you have been foraging in an area.

Fungus flushes will often occur after a combination of rain and mild or warm air temperatures. If a location has not had any rain for a prolonged period (more than 20 days), it will take at least 48 hours before new fungi emerge; if the rain cycle has been regular, then look for new flushes at 12-hour intervals.

Fungi favour edge zones between environments but take care not to forage from areas of high car traffic, historical waste dumps or areas of herbicide use.

Important considerations when foraging for fungi

The key to fungi foraging is to be 100% sure of your identification before collecting and consuming any species. There are species that are deadly toxic when eaten, regardless of the amount. Start slowly and target a few main edible species until you build up the confidence and skills to eat some of the lesser-known ones.

Even with edible types of fungi, when trying a new species you should start out trying only a small amount. Some people will have mild gastric reactions to some fungi, even to porcini, which is recognised as a prime edible in Aotearoa.

Some gilled mushrooms (such as the shaggy parasol and wood blewit), while edible for most people, are known to cause slight to moderate gastric upsets. It is a measured risk as to whether you decide to add these species to your culinary repertoire. *If you are pregnant, or prone to dietary allergies or gastric upset, or are undergoing treatment for serious conditions, it is wise to avoid foraged fungi.*

Identification tips

Learning to take diagnostic photographs of the fungi you find, and the environment from which they grow, is important for accurate identification on either iNaturalist.NZ or the NZ Mushroom Hunting Facebook group (or any other reliable sources you may consult). In some cases, a spore print is also very useful.

To make a spore print, detach the cap of a fresh mushroom and place it gills down on a piece of paper. If the cap is dry, place a drop of water on its top. Cover with a cup or bowl and leave for 2–24 hours, taking note of the colour of the released spores.

Doublecheck the identification features of each fungus as you clean it in the field and place it in a paper bag, then in a container. Take extreme care when picking fungi in the button stage, as species can be easily confused when immature.

For any fungi you are unsure about, take a set of photographs showing:

- The fungi *in situ* — wide angle showing habitat/substrate.
- A close-up showing dorsal view.
- A photo showing the underside — pick one if required.
- A detailed photo of the base of the fungi, either *in situ* or picked. Aim to show all the basal structure even if that means clearing the ground around the fungi or extracting it whole with some substrate attached.
- Show the fungi cut in half in cross-section — if practical, repeat after a 15-minute period to show any colour change in the flesh of the fungi.
- For any fungi collected, place cap-side up over a piece of black-and-white paper with a container over the top of the fungi and leave overnight. Next day, look for a spore print and record it with a digital photograph.

Forage safely

It's critical to avoid picking poisonous mushrooms at all. If a poisonous species is accidentally collected and placed with edibles, it is possible for the spores to cross-contaminate and the whole collection needs to be disposed of.

Never assume that a lot of similar-looking fungi growing close to each other are the same species. Just because a mushroom smells edible, looks attractive or has been eaten by other species such as slugs, *it does not mean that it is safe for human consumption.*

Every so often, new types of fungi arrive in the country, so similar but deadly species may become established and will need to be excluded from your search. Avoid foraging from busy roadsides or areas where herbicides have been used.

Porcini

SCIENTIFIC NAME
Boletus edulis

OTHER NAMES
Ceps, penny bun

DESCRIPTION
One of the easier mushroom species to identify: a stout, white-coloured stem (called the stipe) with a subtle net-link pattern under a brown cap that might have white edges — in other words, no gills but a spore sponge. The spores go from cream-coloured in fresh porcini to yellow-brown in mature specimens. Mature porcini have flatter caps and can weigh over 1kg (porcini is Italian for fat pig).

DISTRIBUTION
Patchy, with strongholds around Dunedin, Christchurch and Wellington. Porcini are most widespread in the Canterbury region, found on lower to mid Canterbury Plains and also at a wide range of sites in eastern North Canterbury. Absent from the upper half of the North Island.

HABITAT
Grows mainly in association with pine or oak trees, but also silver birch, European beech, sweet chestnut and plane trees. Best to look under older trees in mid- to lower-altitude areas.

SEASONALITY
In season from November to July, with a peak from January to May. Occasionally there can be large flushes in December.

FORAGING TIPS
Carefully check under leaf-litter as they are often concealed. Porcini, like many boletes, will re-occur at the same site year after year when conditions are favourable. Porcini have a short shelf-life, so keep cool and dry after collecting, eat fresh when in good condition, and dehydrate, freeze or pickle surplus mushrooms. Store fresh mushrooms separately from older ones.

CAUTIONARY NOTES
Take care to accurately identify, especially when in the button stage, as they can be mistaken for white-gilled *Amanita* species.

Key edible uses

Rich in dietary minerals and vitamin B, this is a very versatile mushroom with a sweet and aromatic scallop-like flavour; one of the most sought-after wild foods in Aotearoa New Zealand. When young and firm (button stage), they are ideal for slicing thinly and sautéing in butter. Large specimens are best dehydrated, which concentrates the flavour. Can be eaten raw when young.

Opposite Fully mature porcini can grow to be huge — as heavy as 1kg!

Peppery bolete

SCIENTIFIC NAME

Chalciporus piperatus

DESCRIPTION

A small bolete with a shiny yellowish-brown colouration above and chestnut-coloured pores below. Typical cap diameter is 2.5–5cm. Under the skin, the stem is bright yellow.

DISTRIBUTION

A very widespread bolete occurring throughout Aotearoa New Zealand, found alongside the (highly toxic) fly agaric, with which it grows symbiotically.

HABITAT

Found in deciduous and coniferous forests, often in association with oak, beech, silver birch, pine and red beech trees, with which it has a mycorrhizal relationship. It can also be found under native trees, especially kānuka, often on the edge of scrub and forests.

SEASONALITY

Mainly February to May, with a peak in April.

CAUTIONARY NOTES

While this mushroom is eaten in many countries, it is extremely peppery so should be used as a flavouring, and not a main meal.

Key edible uses

With a peppery mushroom flavour, it can be sliced up finely and added in small quantities to salads. Also good in stir-fries, though it becomes less peppery when cooked. Usually it is dried, powdered and used instead of pepper as a flavouring.

Birch bolete

SCIENTIFIC NAME
Leccinum scabrum

DESCRIPTION
Has a white-coloured stem with distinct fine black speckling, and a chestnut to yellowish-brown cap. Pores are often a cream colour with a brown spore print. Stem thickness is highly variable, as is the cap colour, depending on which of the four birch bolete types are foraged. Can reach sizes of up to 0.5kg.

DISTRIBUTION
Most common in Southland, Otago, Canterbury, Nelson, Wellington and Taupō regions. Generally rarer in the upper half of the North Island and West Coast.

HABITAT
Only grows in association with silver birch trees, in areas that are not regularly mown or sprayed. Often trees that grow next to waterways will produce larger harvests over a longer season.

SEASONALITY
From November until June, with a peak from January to April.

FORAGING TIPS
Often found in clusters. Note that only about 10% of silver birch trees will have birch boletes growing next to them, so once you find a location it will be worth revisiting to get further harvests.

Key edible uses

Another very versatile mushroom. When young and firm (button stage) they are ideal for slicing thinly and sautéing in butter — delicious placed on a steak. Large ones are best dehydrated to concentrate the flavour. Should be well-cooked to avoid indigestion, and not eaten raw.

Slippery jack group

SCIENTIFIC NAME

Suillus spp.

There are several species of slippery jack that grow only under pine trees in Aotearoa New Zealand, and all are edible. Three species are common, with names that help you distinguish them: the dotted-stalked suillus (*Suillus granulatus*), the purple-veiled slippery jack (*Suillus luteus*), and pungent slippery jack (*Suillus pungens*). There are three other species of slippery jack that also occur under pine trees in this country, but while also edible these are quite rare. Each of these species has slightly different flavours and appearances.

DESCRIPTION

A group of boletes that often grow in tight clusters. Caps are typically yellowish brown, and some types have a distinct mucus covering on the cap; hence the name. The pores on the underside have a yellowish colouration in many types but can also be dark dark cream. Caps are often flatter and narrower than the porcini that may also be found alongside slippery jack species.

The species *Suillus granulatus* has a shiny but not slippery cap and quite bright, yellow-coloured pores; it is the most common slippery jack species in Aotearoa New Zealand.

Suillus luteus is the second-most-common slippery jack in this country. It has a distinct veil covering the pores of immature specimens, which in mature mushrooms recedes to a defined ring on the cap, which is brownish-purple.

The pungent-smelling variety *Sullius pungens* has light yellow-cream pores and an olive-brown slippery cap.

DISTRIBUTION

Found throughout the country in association with pine forests.

HABITAT

Only found under pine forests. Can occur under both plantation and wild pines.

SEASONALITY

A long season from November to August, with a peak from February to June.

Opposite Slippery jack grow only under pine trees in this country, as you can see by the surrounding pine needles.

FORAGING TIPS

Generally found in larger amounts under older pine trees. Often found along the disturbed ground on the edge of pines. Only about a quarter of pine trees will have slippery jacks growing with them, and they also tend to be very patchy in their distribution. However, large numbers can be harvested, especially in the late autumn and early winter.

CAUTIONARY NOTES/CONFUSION SPECIES

To check you haven't picked one of the inedible bolete species, bruise the cap's underside — if it stains the paper blue, do not use it. The skin can cause gastric discomfort, so peeling older specimens in particular is recommended. Do not eat raw.

Key edible uses

Can be eaten fresh in the early button stage. Larger ones should be peeled and are best dehydrated as they can have an unappealing soft, moist texture when cooked. This is an ideal mushroom to dehydrate then crush into a powder to use as a seasoning blended with salt. Popular also for pickling when in the button stage. Can also be smoked.

Inset Cap of dotted-stalked Suillus, *Suillus granulatus* **Above opposite** Underside of pungent slippery jack *Suillus pungens* **Below** Underside and side-view of purple-veiled slippery jack, *Suillus leteus.*

Larch bolete

SCIENTIFIC NAME
Suillus grevillei

OTHER NAME
Greville's bolete

DESCRIPTION
A small to medium-sized bolete with a yellowish-brown cap and bright-yellow pores, and often a short stalk. Has a shiny cap, sometimes with a light mucus covering.

DISTRIBUTION
Found mainly in mid-altitude areas of the South Island and some higher-altitude parts of the North Island.

HABITAT
Only grows next to larch trees, which along with a mossy groundcover offer the best foraging options. Alternatively, check sloping banks, as large boletes like growing from slightly disturbed soils on edge environments (especially of groves of larch trees and to a lesser extent on larch tree plantations). They often appear in clusters but can be well hidden among the larch needles.

SEASONALITY
From November to May, with a peak in February to April.

FORAGING TIPS
Often occurs in clusters.

CAUTIONARY NOTES
Take care to correctly identify, especially in the button stage, as similar-looking but toxic gilled mushrooms will often occur next to larch bolete clusters.

As with the slippery jacks (see page 436), the skin can have purgative properties, so peeling older specimens is recommended. Do not eat raw.

Key edible uses

Best eaten when small buttons, which can be either fried or pickled. Larger mushrooms should be peeled. They have a high moisture content but are suited to dehydrating and have an aromatic citrus flavour.

Inset A mature larch bolete, showing its porous underside. **Opposite** Larch boletes often have an orange tinge, but can also be yellow.

Western painted suillus

SCIENTIFIC NAME
Suillus lakei

DESCRIPTION
Characterised by the texture of tufted fibres or small scales on its reddish-brown cap. It can also be identified by the woolly-looking veil on its stem. The pores are a distinct yellow honeycomb, bruising brown when touched.

DISTRIBUTION
Found throughout Aotearoa New Zealand in association with Douglas fir trees, but tends to be more common in the South Island, especially Central Otago, preferring mid-altitude areas.

HABITAT
Only found underneath Douglas fir trees.

SEASONALITY
A late-season fungus typically found from February to the end of May.

CONFUSION SPECIES
If there are no Douglas firs where the mushroom is growing, it will be a different species and should be avoided.

Inset Underside of a picked western painted bolete with fresh growth Douglas fir. **Opposite** Note the distinct rough-textured reddish-brown cap.

Key edible uses

A good eating mushroom when finely sliced and fried. Quite a fleshy bolete, it is ideal for dehydrating. Tests on this species in Mexico showed antimicrobial properties. Do not eat raw.

Horse mushroom

SCIENTIFIC NAME

Agaricus arvensis

DESCRIPTION

A medium to large-sized white mushroom, usually bigger than field mushrooms (see page 446). In the button stage it tends to have a steeply shaped cap that flattens out as the mushroom matures. Gills are pale pink maturing to dark brown, with a brown spore print. There is a partial veil covering the inner gills, which will later drop to form an untidy skirt. Occasionally will have a very slight yellow discolouration on stem — not as extreme yellow as yellow stainer (see page 480), but if in doubt, don't use. Smells sweetly of almonds or aniseed.

DISTRIBUTION

Widespread around Aotearoa New Zealand.

HABITAT

Likes rough farmland and large open grass areas such as parks and grass verges.

SEASONALITY

Main season is from November to May.

FORAGING TIPS

Best to gather small to medium-sized specimens. Leave the large mushrooms to continue spreading spores. Often grows in mushroom circles known as 'fairy rings'.

CAUTIONARY NOTES

See the notes for field mushroom (see page 446) — in particular take care not to confuse with the yellow stainer mushroom, *Agaricus xanthodermus* (see page 480), which looks similar but will have a vivid yellow colour and acrid 'chemical' smell.

Key edible uses

A top eating mushroom that can be used in a wide variety of ways. Often fried, baked or grilled. Horse mushrooms can be foraged in large amounts and tend to have flushes. At such times, stewing the mushrooms and storing in jars is a useful option. Young specimens may be eaten raw.

Inset A mid-stage horse mushroom with its distinctive rough-textured cap. **Opposite** Freshly foraged horse mushrooms in prime condition. Don't confuse these for the toxic yellow stainer mushroom.

Field mushroom

SCIENTIFIC NAME

Agaricus campestris

OTHER NAME

Meadow mushroom

DESCRIPTION

A small to medium-sized white mushroom. The button stage tends to have a rounded cap that flattens out as the mushroom matures. Gills begin pale pink, maturing to dark brown, almost black, creating a brown spore print. Occasionally the stem will have a very slight yellow discolouration. Has a pleasant aromatic smell. Smaller in size than the horse mushroom (see page 444), its largest specimens have a cap diameter up to 15cm wide.

DISTRIBUTION

Throughout Aotearoa New Zealand.

HABITAT

In areas with short grass and pasture. Tends to be absent in soils that have been modified by modern agricultural practices.

SEASONALITY

Main season is from November to May, with a peak from February to April.

FORAGING TIPS

Stew when collected in large amounts, or dehydrate or pickle. It is a fragile mushroom, so store carefully in containers.

CAUTIONARY NOTES

Take care to accurately identify; check for a brown spore print. The bruised cap should produce a light-pink stain; if it turns bright yellow, it is likely to be *Agaricus xanthodermus* (yellow stainer; see page 480), which is poisonous. The yellow stainer might also have an acrid smell. Be careful not to confuse with white-gilled *Amanita* (death cap; see page 476) or rust-gilled *Cortinarius* mushrooms (see page 480), especially when foraging near trees. The death cap grows from a distinctive base like an extra cylinder around the bottom of the stem, known as a volva; the field mushroom does not. *Cortinarius* species are mostly purple so easy to avoid, but if your mushroom cap folds under into a kind of pouch, covering where you'd find gills in a field mushroom, then discard.

Key edible uses

This is a top eating mushroom, which can be used in a wide variety of ways. Often pan-fried, baked or grilled. Can also be pickled in the button stage.

Honey mushroom

SCIENTIFIC NAME
Armillaria novae-zelandiae

MĀORI NAME
Harore

DESCRIPTION
Has a honey-coloured cap with a light mucus covering; the cap is 4–15cm in diameter with firm white flesh. It has cream-white gills and a white spore print. Grows in clusters from a central base and has a darker brown stem at the base, which lightens in colour towards the cap. It has a ring on the stem and might give off a weak bioluminescence at night. (NOTE — Take extreme care when foraging as it can easily be confused with the deadly poisonous funeral bell.)

DISTRIBUTION
Throughout Aotearoa New Zealand but tends to be more common in lowland areas that are relatively frost-free. Especially common on the South Island's west coast.

HABITAT
Grows on dead wood, primarily in softwood native forests. Tends to be more common in native forests with lots of decaying logs.

SEASONALITY
Can be foraged from December to September, with the main season being March to June with peak harvests in May. Dies away in spring and out of season in October to November.

FORAGING TIPS
Grows in clusters on dead wood. Pick small and medium-sized ones, leaving larger specimens to continue sporing. They have a short shelf-life so either eat or dehydrate promptly after harvest. Best put upright in a cardboard box with paper towels when harvested, otherwise grit can easily accumulate in the gills.

CAUTIONARY NOTES
Take extreme care not to confuse with the funeral bell, *Galerina marginata*, which can also favour dead wood. Unlike the honey mushroom, the gills of the funeral bell begin pale brown and darken on maturity. Other white-gilled mushrooms to be wary of are the *Amanita* genus, such as the death cap (see page 476).

Key edible uses

This traditional food source for Māori is best eaten fresh and stir-fried or cooked in seasoned butter. It has a mild umami flavour with a high water content, making it good to add to soups. *Armillaria limone* and other species of this fungus are bitter and best avoided. This mushroom must be well-cooked.

Opposite Fully developed honey mushrooms grow in clusters.

Shaggy parasol

SCIENTIFIC NAME

Chlorophyllum rhacodes

DESCRIPTION

A medium to large, white-gilled mushroom with distinct scaly brown markings on its delicate cap, which begins rounded before flattening out. When cut in cross-section it develops a reddish-brown colouration. With a definite ring or skirt on its stem, it often grows in clusters. Spores are white.

DISTRIBUTION

Widespread in lower-altitude areas with strongholds in Canterbury and Otago.

HABITAT

Tends to like areas of disturbed soils, leaf-litter underneath trees and nutrient-rich soils. Often found growing in association with macrocarpa and pine trees and a wide range of larger native trees.

SEASONALITY

From October to August, with a peak from December to May.

CAUTIONARY NOTES

Because it also has white gills and spores, take care not to confuse with *Amanita* mushrooms (see page 476), especially when the mushroom is in the younger button stage. Also cut in cross-section and check for red colouration to distinguish from other parasol species. While not common in Aotearoa New Zealand, shaggy parasol can also be confused with the toxic dapperling mushroom, *Lepiota* spp.

Key edible uses

The cap is mainly eaten, as the stem is quite fibrous. It has a firm texture and rich flavour. The cap of larger mushrooms is ideal for serving covered with melted cheese and a sprinkling of herbs. Some people experience an allergic reaction to this species, so test before eating in quantity.

Inset Shaggy parasols have distinct scaly, shaggy caps. **Clockwise from left** Showing distinct cap markings; a cluster of fresh buttons; a fully developed cap, with a fragile structure and white gills. Take extreme care to correctly identify this species, as its easily confused with the infamous death cap.

Wood blewit

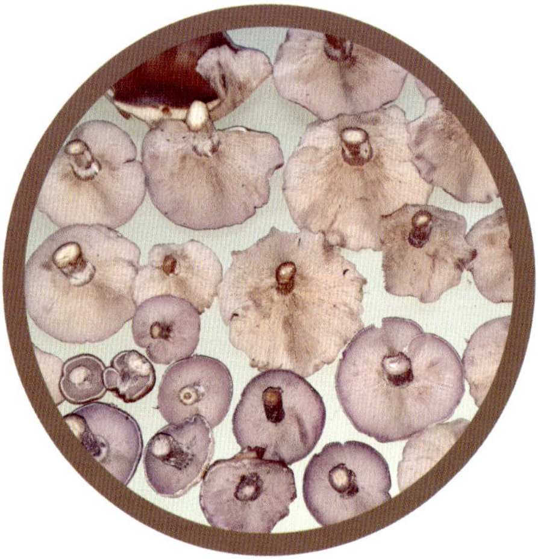

SCIENTIFIC NAME
Clitocybe nuda

OTHER NAME
Lepista nuda

DESCRIPTION
Has distinct lilac-coloured gills and a light-pinkish cap. Its cream to pale-pink spore print is important for identification. It has a thick, fibrous, white to pale-purple stem (with no brown streaking).

DISTRIBUTION
As an introduced species, its spores have mainly spread around the key urban centres of Dunedin, Christchurch, Wellington and Auckland plus some regional centres (such as Timaru and Palmerston North), but it is likely that the range and abundance of this mushroom will increase.

HABITAT
A spore-spread mushroom that grows in among leaf-litter of a wide range of trees. Often found in pine forests, but also occurs in native bush, generally in areas with good light. Also found in and around urban parks, along forest edges and tracks. May be found in clusters.

SEASONALITY
Mainly found from April to July, with a peak in May and June.

CAUTIONARY NOTES
Care needs to be taken with accurate identification. Can cause gastric upsets in a small number of people, so try a small amount first and make sure it is well cooked. Take care not to confuse with some of the poisonous purple-coloured *Cortinarius* species (which have brown spore prints, and when immature often a veil hanging under the cap). Avoid using raw, as it can cause allergic reactions.

Key edible uses

A top mushroom to cook whole, with a firm texture, good flavour and slight citrus aroma.

Inset A harvest of mature wood blewits. **Above opposite** Fresh growth wood blewits in a range of stages. **Below** Fully mature wood blewit.

Shaggy inkcap

SCIENTIFIC NAME
Coprinus comatus

DESCRIPTION
A white mushroom that starts out with an elliptical cap (like an upright pointed egg), which is shaggy and hugging its thick short stalk, and as it matures opens out to a typical umbrella shape. The gill colour also changes from white to pink to jet inky black as the mushroom ages, with the mushroom ultimately breaking down into a black liquid — hence the common name.

DISTRIBUTION
Throughout Aotearoa New Zealand but rare north of Auckland and in west coast regions of the South Island. Generally found in open lowland areas but occurs at inland sites in Central Otago.

HABITAT
Likes well-drained and disturbed soils. Often found on the edge of gravel tracks, and usually avoids damp areas with lots of grass.

SEASONALITY
Most common from February to early May.

FORAGING TIPS
Often grows in clusters so look carefully in the surrounding area when you find one. Once gathered, place in a container to prevent crushing, keep cool, and use as soon as possible.

CAUTIONARY NOTES
Although only a recent migrant to the country, the common inkcap, *Coprinopsis atramentaria*, is an issue as it has a similar changing shape. It is edible, but poisonous when combined with alcohol. Another recent import with the similar egg-to-bell-to-plate stages is mica cap, *Coprinellus micaceus*, which is in fact edible and high in potassium.

Key edible uses

Fresh shaggy inkcaps are firm so can be finely sliced lengthways and fried in butter. The cap has a very strong field mushroom flavour. The stalks can be quite fibrous, and many choose to discard them. When mature, the inkcap is soft and produces a black liquid that can be collected, ideally slowly in the fridge, and used to colour and flavour dishes such as soups, stews and risottos, in place of squid ink.

Opposite Fully mature shaggy inkcap.

Tawaka

SCIENTIFIC NAME
Cyclocybe parasitica

OTHER NAME
Agrocybe parasitica

DESCRIPTION
A large white to buff-coloured mushroom with pink to dark gills and a white stem that often has a distinct white veil that covers the gills and recedes to form a distinctive skirt or collar on the stem, which appears brown when dusted with spores. It looks like a field mushroom with a firm stalk, but is usually seen growing in clusters on trees, the stems linking to a central base. Has a dark-brown spore print.

DISTRIBUTION
Widespread throughout the North Island except for the Central Plateau. Found on the eastern side of the South Island in coastal areas, but a lot more common north of Banks Peninsula. Also found in Nelson and Golden Bay regions but largely absent from the South Island's west coast. Primarily has a coastal-lowland distribution but can be found at inland sites.

HABITAT
Found on dead and dying wood from a wide range of host trees, but generally not on conifers.

SEASONALITY
From November until June, with a peak from January to May. Small numbers can be found throughout the year, especially in the upper half of the North Island.

FORAGING TIPS
Often occurs in small clusters. Pick the fresher-growth mushrooms, leaving the larger ones to spore. Will often regrow at the same site. Use a sharp knife to cut off the tree.

CAUTIONARY NOTES
Care needs to be taken with accurate identification. Could be confused with other mushrooms that grow out of decaying wood, such as the more-orange spectacular rustgill, *Gymnopilus junonius* (see page 479).

Key edible uses

Used for traditional Māori medicine, this is a top eating mushroom with a strong flavour. Best fried, but can also be dehydrated and made into a seasoning.

Opposite Mature tawaka, and of the fungi's main host trees, the lacebark.

Velvet shank

SCIENTIFIC NAME

Flammulina velutipes

OTHER NAME

Velvet foot

DESCRIPTION

A small mushroom with a cap up to 5cm in diameter, though typically 2–3.5cm. It grows in tight clusters on tree trunks or piles of decaying wood, starting as buttons and opening in stages out to dinner plates. The stem is pale pink maturing to black and has a distinct velvet-like appearance. The cap is light orange or straw-coloured and often has a slight mucus covering. It has white gills that become creamy on maturity and its spore print is white.

DISTRIBUTION

Found throughout lowland Aotearoa New Zealand but with patchy distribution at low- to-mid-elevation sites. As a recently introduced species, it is likely that it is still increasing in distribution and abundance.

HABITAT

Often grows in damp bush, on decaying softwood trees and dead wood (but not conifers). Commonly found on the edge of forest patches and along walking tracks, suggesting that people are helping the spores to spread.

SEASONALITY

In season primarily from April to September with a peak from June to August, and considered a winter mushroom.

FORAGING TIPS

Take care when gathering to brush dirt off the mushrooms, and keep upright as grit can easily accumulate in the gills. The caps can be gently wiped with a paper towel to remove grit and its light mucus layer. They have a moderate moisture content so are best eaten within a few days of harvest. When collecting leave some clumps, especially of older mushrooms, so that the spores can spread. Often will be found regrowing at the same site.

CAUTIONARY NOTES

Take extreme care not to confuse with the funeral bell, *Galerina marginata* (see page 479) or the white-gilled *Amanita* genus (see page 476).

Key edible uses

A small, delicate but sweetly flavoured mushroom that is suited to using as a garnish. Best lightly sautéed with butter or added to soups.

Opposite Velvet shank mushrooms in a range of growth stages. **Inset** This fungi typically grows in clusters on dead and dying wood.

Fairy ring champignon

SCIENTIFIC NAME
Marasmius oreades

OTHER NAME
Scotch bonnet

DESCRIPTION
A small mushroom that typically grows in circles in short grassland. Cap diameters are from 1cm to 6cm. Has a distinct straw-coloured cap and a cream-coloured spore print. The cap usually has a raised lump in its centre. The white or creamy gills do not connect to the stem, and the stalk is white and fibrous. The combination of these factors allows the fairy ring to be identified.

DISTRIBUTION
Found throughout Aotearoa New Zealand, preferring drier regions.

HABITAT
Often found on road and lawn edges and on drier farmland with low grass. Most common in areas where the soil structure is intact, and often grows in circles among yarrow and dandelions. The fungi like good sunlight and reasonably well-drained soils but do need some moisture to set off flushes.

SEASONALITY
October to May.

FORAGING TIPS
Often found in clusters and large flushes can occur, so it pays to make the most of harvesting when available. Typically, will grow in rings. It is best to cut the stem halfway down, which helps with the handling of the mushroom.

CAUTIONARY NOTES
Take care to identify each mushroom separately and note the identifying details in the description to distinguish it from any imposters.

Key edible uses

A very versatile mushroom with a firm texture and sweet nutty flavour. For this reason it can be used with desserts, and it makes a good garnish mushroom too.

Clockwise from above A fully mature mushroom; typical habitat growing in a distinctive fairy ring on a lawn; freshly foraged mushrooms.

Brown oyster mushroom

SCIENTIFIC NAME

Pleurotus australis

DESCRIPTION

A greyish-brown mushroom from above with cream-coloured gills and a very short, thick-set base. Grows to 10cm high either singly or in clusters from tree trunks and fallen logs. The cap begins as a button, becoming an umbrella shape before flattening out, then may invert into a funnel shape.

DISTRIBUTION

Most common in the upper half of the North Island but also found in Wellington and Nelson-Golden Bay regions. Largely absent from much of the South Island.

HABITAT

Found on dead and decaying wood in warm forest environments.

SEASONALITY

Can be found throughout most of the year with a summer to autumn flush and a smaller spring flush. Tends to die away in late winter to early spring.

FORAGING TIPS

Take care to remove from the dead wood with a sharp knife and keep upright to avoid any dirt getting into the gills.

SIMILAR SPECIES

The edible native purple oyster mushroom (*Pleurotus purpureo-olivaceus*) is a smaller oyster mushroom, mostly flatter, though it may also invert into a dish or funnel shape. The amount of light available determines its colour, which could be dark purple to brown to greeny-yellow. The stems are white to yellow, while the gills are cream with a light-purple tinge. Like the brown variety, these fungi cluster on wood but are more separated and usually found in beech forests.

Key edible uses

Oyster mushrooms are rich in vitamins, minerals and antioxidants. A robust but small mushroom with a mild flavour, best suited to frying.

Opposite A cluster of fresh growth brown oyster mushrooms growing on a decaying softwood log.

Wine-cap stropharia

SCIENTIFIC NAME
Stropharia rugosoannulata

OTHER NAMES
Burgundy mushroom, burgundy wine cap, king stropharia

DESCRIPTION
A medium-sized gilled mushroom with a distinct burgundy cap, which lightens as the mushroom grows and its rounded shape flattens out. The gills are grey to greyish-black, producing purple-grey to purple-black spores.

DISTRIBUTION
Most common in Auckland, Waikato regions and Bay of Plenty, but also found along Kāpiti Coast. More common in the North Island but also found in the south with strongholds around Christchurch. It appears to be slowly spreading in distribution, with pockets found at random locations throughout the country, often where accidentally inoculated wood mulch has been distributed around roadside environments.

HABITAT
Can grow on gravel riverbeds close to the water, and typically found in areas of disturbed soils in and among shrubs and garden edges. They tend to like areas with wood chips and decaying matter.

SEASONALITY
Has both a spring and an autumn flush, found less often in summer and dying away in winter.

FORAGING TIPS
Often found in clusters, so carefully check for more mushrooms in close proximity.

CONFUSION SPECIES
Take extreme care when foraging any type of fungus to correctly identify beyond doubt. As they occur in clusters, also take the time to confirm the identity of each mushroom picked. If it has white gills, leave well alone as it is likely to be a death cap (see page 476).

Key edible uses

Cooked and used in the same way as the field mushroom, it has a mild flavour and is suited to frying or baking. Buttons can be pickled.

Above opposite A harvest of mushrooms in prime condition. **Below** Fully mature mushrooms growing among woodchips on a bank. **Insert** A button, showing the deep burgundy cap.

Wood ear fungus

SCIENTIFIC NAME
Auricularia cornea

MĀORI NAMES
Hakeka, hakeke

OTHER NAMES
Cloud ear, ear fungus

DESCRIPTION
A type of bracket fungus with no gills but a hairy surface and rubber-like texture. Initially the fungus is a uniform dark brown, but lightens as it matures and has a whitish colouration when sporing. It grows in tight clusters on dead wood.

DISTRIBUTION
Lowland forests throughout the country, but more common in northern South Island and throughout lowland North Island. Scarce in the lower half of the South Island.

HABITAT
Generally found in softwood lowland forests growing on a wide range of dead hardwoods. Likes shaded locations with mild temperate and damp conditions.

SEASONALITY
Available year-round but has peak growth over the winter months.

FORAGING TIPS
Fresh growth is the best to forage, leaving the older, harder and drier fungi to spread their spores.

Key edible uses

Mildly flavoured with a good crunchy texture, this fungus is commonly included in Chinese dishes. It can be eaten raw when finely sliced but is best stir-fried or steamed, which will slightly soften the flesh and is closest to the hāngī, which was how it was cooked by early Māori. It absorbs flavours well, making it useful in broths and soups. It can be dried.

Opposite With its distinct form and texture this is an easy fungi to identify.

Turkey tail

SCIENTIFIC NAME
Trametes versicolor

OTHER NAME
Bracket fungus

DESCRIPTION
A thin bracket fungus with a series of rings in brown, purple and white on top and a cream-white colouration and white spores below. Often grows in dense clusters.

DISTRIBUTION
Throughout Aotearoa New Zealand, except for very-high-rainfall or high-altitude areas.

HABITAT
In forests and damp shaded areas, found growing near the forest floor on decaying logs and branches.

SEASONALITY
Available throughout the year but peaks over the cooler months from April to September.

FORAGING TIPS
Use a sharp knife to collect the specimens with the most vivid colours, avoiding the brown and dried-up ones or those with green algae growing in them. Leave the older ones to continue spreading spores.

SIMILAR SPECIES
Turkey tail is easily confused for *Stereum ostrea*, a non-toxic fungus. *Stereum ostrea* does not have hairs on the upper side, and the spore layer is white, not cream.

Key edible uses

Mild woody flavour with a slightly bitter aftertaste. It is too tough to eat, and is usually steeped in hot water to make a pleasantly woody-flavoured golden tea. Used in Chinese medicine, its potential medicinal properties have been the subject of much recent research, along with its reported antiviral and antioxidant activity.

Clockwise from left Turkey tail grows in clusters; fully matured turkey tail, showing striation; prime, fresh growth to pick; with unique white edges.

Giant puffball

SCIENTIFIC NAME
Calvatia gigantea

MĀORI NAME
Pukurau

DESCRIPTION
A round or elliptical white fungus that ranges from 10cm to 60cm in diameter and can weigh up to 10kg. A cream-white coloured outer membrane encircles the puffball. When fresh (and edible), the flesh is a pure white colour. Once there is any brown colouration inside, the puffball is starting to develop spores and is considered inedible. Spore print is a yellow-brown colour.

DISTRIBUTION
Throughout the country except in higher-rainfall forested areas, but most common in Southland, Otago and Canterbury. Tends to be less common in the North Island.

HABITAT
Being a spore-spread fungus, it can turn up in a wide range of edge environments, even on the side of urban walkways and gardens. But it is most common in areas of rough pasture with minimal agricultural intensification.

SEASONALITY
Mainly October to May, with a peak from February to April.

FORAGING TIPS
Often grows in clusters or circles with the puffballs at different stages. Leave any discoloured (brown) puffballs to produce spores. When a large puffball is found, it is worth looking in the long grass nearby for others. After harvesting, keep in a low-moisture environment.

CAUTIONARY NOTES/CONFUSION SPECIES
Take care when collecting to check that the flesh is clean and white and not starting to develop spores. Avoid collecting from nutrient-rich soils that are polluted. Other fungi (such as death caps; see page 476) can look like puffballs when they emerge, so cut through to check it is completely flesh with no stem or separate cap forming. Brown, black or dark-purple internal flesh means you have picked an earthball (*Scleroderma* spp; see page 481), which is poisonous. Still be cautious if it looks like a potato but has a white centre as this might be a young false truffle, a member of the *Rhizopogon* species, which will have a dark interior as it matures, and although it is unlikely to poison you it is not worth eating.

Key edible uses

Another fungus used in traditional Māori medicine, it can be eaten at any size as long as the flesh is pure white throughout. The firm flesh has a mild field mushroom flavour and is very versatile. Can be used like bread or a pizza base and lightly fried or baked. Also can be cut into cubes and added to soups. Ideal to dehydrate.

Pekepeke-kiore

SCIENTIFIC NAME
Hericium novae-zealandiae

OTHER NAMES
Fungus icicles, lion's mane

DESCRIPTION
Grows as a cluster with fine, white, icicle-like growths facing downwards. This distinct and sought-after fungus has a maximum size of 80cm in length. As it matures and produces spores, it starts to discolour to off-white.

DISTRIBUTION
Scattered throughout Aotearoa New Zealand but tends to be more common in the North Island and in low- to mid-altitude parts.

HABITAT
Mainly in native forests but occasionally in the undergrowth of old pine forests. Tends to grow in decaying trees, including inside decaying tree trunks.

SEASONALITY
Available all year, but main season is December to June.

FORAGING TIPS
When foraging, leave some of this growing on the tree to allow it to mature and allow dispersal of spores. As it is naturally rare, foragers are encouraged to try home cultivation of this species.

Inset This pretty fungi grows from crevices in decaying wood.

CAUTIONARY NOTES
Only consume when fresh with clean white flesh.

Key edible uses

Lightly pan-fried or added to stir-fries. It has a mild sweet flavour that has been likened to some crustaceans. Research has indicated that this fungus contains beneficial properties.

Landscaping black morel

SCIENTIFIC NAME
Morchella importuna

DESCRIPTION
Only recently identified in this country, this black morel has a distinct dark indented ripple pattern on its bell-shaped cap and a hollow white stem as a base.

DISTRIBUTION
At sites in both North and South Islands near urban centres, but as it becomes more recognised is likely to increase in number and spread.

HABITAT
Often in wood mulch and areas of disturbed soils.

SEASONALITY
A spring-time fungus found mainly from September to November.

FORAGING TIPS
Often well-camouflaged among wood chips and mulch, so you need to carefully look for them. Often occurs in clusters and although a spore-spread species it will re-occur at the same site between years.

CAUTIONARY NOTES
Cook this fungus, as it is poisonous raw. Do not confuse with the poisonous false morel, *Gyromitra* spp. — these are rounder and folded (more brain-like) and their stems are not hollow.

Key edible uses

One of the most sought-after of edible fungi, it has a sweet, delicate and aromatic woody flavour. It needs to be cooked, not eaten raw.

Inset The morel mushroom can often be well-camouflaged, so look for this fungi carefully.

Reindeer moss

SCIENTIFIC NAME
Cladonia confusa

OTHER NAME
Fuzzy reindeer lichen

DESCRIPTION
Despite its name, this is a lichen, a symbiotic partnership of both fungi and algae. The *Cladonia* is a genus mostly eaten by reindeer. It is an intricately structured green-white lichen that grows in mats on the ground and often in mosaics with other lichen species. It has a fine form with a crunchy texture.

DISTRIBUTION
Throughout Aotearoa New Zealand but generally in higher-altitude areas, above 300m, along the Southern Alps and the Central Plateau, but will occur in other sites at lower altitudes in exposed areas.

HABITAT
Damp alpine basins, and often grows in association with mosses. Generally found in open environments or along the edge of shrubs but likes damp basin areas of flat ground.

SEASONALITY
A slow-growing species, it is available throughout the year.

FORAGING TIPS
Do not gather lichen in polluted areas, and wash thoroughly before use. As it has very patchy distribution and is slow-growing, harvest in small amounts. Use a knife to cut off the upper half of the lichen as the base is often soft and full of leaf-litter. Lay flat in a tray once harvested to avoid particles contaminating the reindeer moss.

CAUTIONARY NOTES
Cladonia is an acidic species that may cause gastric upset. This can be reduced if the lichen is first boiled in water with baking soda for a quarter of an hour, then strained and boiled up again.

There are at least 29 native species of *Usnea*, otherwise known as beard lichens, plus several other lichens with *usnea* as their second name. They are nearly all very similar tangles of greyish filaments — looser and longer than *Cladonia* — usually growing off trees, and the species tends to share the feature of an inner elastic that is revealed when you break open a filament (as in chickweed, *Stellaria media*). A few *Usnea* species are browny-red or yellow, some longer or thicker than others. Some people grind the filaments into flour for baking and traditional medicine.

Opposite Reindeer moss is not a fungi, but a lichen, and got its name from being a main food source for reindeer.

Key edible uses

Esteemed by chefs as a creative plating device on which to serve gourmet foods, it can, however, be eaten and has a mild, earthy flavour. It should be cooked: frying will make it crunchy, while pressure-cooking will soften it. It is tasty sprinkled with ground dehydrated mushrooms. Lichen can be used to make tea and was a staple ingredient in aquavit.

Poisonous Fungi

It's crucial to be 100% sure of your identification before collecting and consuming any mushroom species. Remember to start slowly: researching the key edible species before heading out, foraging only for those ones; identify each species as you pick it and store each type of fungi separately once collected.

There are no golden rules for determining whether a fungi is edible or poisonous. For helpful fungi identification tips, see page 431.

Fly agaric, *Amanita* spp. ⚠

The most recognisable poisonous fungus is the *Amanita muscaria*, identified by its red colour and white spots. The fly agaric is in the *Amanita* genus, which contains a range of toxic species, so is worth recognising and avoiding. Amanitas are characterised by clean white gills — take extreme care to accurately identify white-gilled mushrooms. Some have warty caps. There are some edible amanitas but they require a high level of expertise to identify.

Death cap, *Amanita phalloides* ⚠

The death cap is the deadliest fungus and accounts for the majority of fungi deaths globally. Death caps are found in the Nelson region and upper half of the North Island, but can spread. They grow in association with Oak tree roots. The gills are white and produce a white spore print, which will distinguish it from similar fungi you forage such as wine-cap stropharia (see page 464). The colour of the cap can vary, making it tricky to identify but the key to identifying the death cap is the presence of a volva at the base. It emerges as an oval covered in a white veil, which can make it look like a puffball, and which is often still visible at the base of the mature stalk as a ragged sheath. The oval flattens into a cap with tones of green, yellow or orange. It has a sweet fragrance that turns sickly as the fungi ages.

Jewelled amanita, *Amanita junquillea* ⚠

There is a range of other poisonous mycorrhizal *Amanita* species in Aotearoa New Zealand, with the jewelled amanita (or gemmed amanita) being common in pine forests around Wellington and often encountered by foragers looking for porcini. They can be confused with porcini in their button stage, but this fungi has white gills and a white spore print, which differs from porcini. The yellowish caps of jewelled amanita flatten out and have warty 'gem' markings on the surface; they may not be numerous or as obvious as other warty amanitas. Its volva becomes a ring at the very base of the stalk.

Marbled death cap, *Amanita marmorata* ⚠

Another common *Amanita* species that is deadly poisonous is the marbled death cap. Note that it also has the clean white gills that are characteristic of the *Amanita* genus. As its common name suggests, it has a grey-brown, slightly marbled, almost streaked, cap when mature. Again there is a noticeable volva at the base. It has a strong aroma.

False blushing amanita, *Amanita excelsa*, and Grey-spotted amanita, *Amanita excelsa* var. *spissa* ⚠

Some online sources say that these fungi are safe to consume and should only be avoided because of their similarity to a fungus that has not been naturalised in Aotearoa New Zealand. However, there are cases recorded where variants have caused severe kidney damage. It may be that the toxins are destroyed by heating, but do not take any risks with this fungus until further research has been done. Mainly found in the Canterbury region, favouring pine forests and oak trees, its grey-brown warty cap makes it easily recognisable. The gills are white, the stout white stem often shows its veil, and a well-defined ring divides the stem into a smooth top and a scaly bottom. Its unpleasant aroma suggests it is best avoided.

Mulch Fieldcap, *Agrocybe putaminum* ⚠

A very common gilled mushroom found mostly in the South Island growing in wood chips, sometimes in fairy rings, spread by spores. It has a pale-cream to yellowish cap, which may be slightly darker in the middle. The young form has a veil that partially covers the gills, which are off-white-cream becoming browner as it matures. In addition, in moist conditions the gills can look as though they are covered with brown spots as they accumulate water droplets that leave marks when dry. The spores produce an orange-brown print. The yellow-white stem swells slightly at its base and is faintly mottled at its top. It has a smell reminiscent of radish and a bitter flavour. While poisonous, it is more likely to give you abdominal pain and make you vomit than kill you.

Funeral bell, *Galerina marginata* ⚠

A deadly poisonous, small brown mushroom that grows on decaying wood, particularly conifers. Extreme care needs to be taken when foraging honey mushroom species (see page 448) and velvet shank mushrooms (see page 458) as these both look similar to the funeral bell, which has the same lethal compounds as those found in the death cap mushroom (see page 476). The cap of the funeral bell is yellow-brown, becoming slightly paler at the outer edges. The caps begin curved like an umbrella but with age widen out more like a parasol, even becoming flat, often with a raised knob in the centre. The gills are cream to honey-coloured and produce an orange spore print. Immature specimens still have a ring of membrane towards the top of their whitish stem, which is darker below the ring and becomes hollow as it matures.

Spectacular rustgill, *Gymnopilus junonius* ⚠

Also known as laughing gym, this is a very common mushroom that grows in large quantities in tight bunches on dead wood. Often reaching an impressive size — larger than a hand — they easily attract the attention of foragers, but it is a toxic species. The young buttons become parasol-shaped caps, then dinner plates, which are varying shades of browny-orange with scaly markings and often lighter margins. As the name suggests, the spore print is rusty, coming from gills that start out yellow and become rustier. The stem is yellow and fibrous below the ring. Take care when foraging tawaka (see page 456) so as not to confuse it with the spectacular rustgill.

Yellow stainer, the Xanthodermatei ⚠

A grouping of gilled mushrooms, including *Agaricus xanthodermus* in the field mushroom genus, which could be easily confused with a range of edible field mushroom species, given the white cap and gills that are pale pink darkening to brown with age. The spores are brown. However, they have a pungent, chemical-like smell and when cut in cross-section will immediately stain a bright-yellow colour at the bottom of the usually bulbous stem and sometimes in the cap. This colour will turn a whitish gray. While not a deadly poisonous species, they can induce severe vomiting and are colloquially known as the 'vomit comet'.

Webcaps, *Cortinarius* spp. ⚠

A very diverse group of gilled mushrooms, many of which are toxic. Some of the purple-coloured species could easily be confused with the edible wood blewit (see page 452). Known as 'corts' in the fungi foraging fraternity, referring to the cortina or veil that covers the gills when the fungi are buttons, this distinguishing element starts to detach and become web-like (hence the common name) as the fungi grow but unfortunately disappears on maturity. They come in a range of — often very distinctive — colours: purples and blues, greens and yellows, as well as greys and browns. The spores are generally orange-red-brown. With over 150 species in this country, it is best to admire but leave alone.

Earthball, *Scleroderma* spp. ⚠

A very common grouping of eight species that could be confused with a range of edible button-staged mushrooms and puffball species. The earthball, however, has a deep brown, black or dark-purple colouration inside (note that puffballs should only be eaten when they have clean white flesh throughout when sliced). Rounded in shape, growing on a short stalk, earthballs may have warty markings and split and peel open when they spore to reveal an unappealing dark centre. The potato earthball, *Scleroderma bovista*, is the most widespread earthball in Aotearoa New Zealand, and as its name suggests it looks like a potato.

False truffles can also look like potatoes, as does the violet potato fungus, *Gallacea scleroderma*, and there are other fungi species that are similar in being ball-like, which are best to avoid — if they are not all poisonous they are certainly not palatable. They might open up like stars, be marked by orange lips on top, look like stones, be mottled or bright orange or purple. Mostly they will have dark insides or insides that look like a walnut kernel. Leave well alone.

Recipe Basics

Foraged items are generally highly nutritious, vitamin-rich and a source of such minerals as iron, calcium and magnesium. The following recipes are basic, and you may well find more interesting variations and different plants to add — these are merely to suggest a starting place and offer inspiration for the sorts of things you can do with your foraging. Note that as there will be different water and sugar content in the different species you collect, some of the ingredients or cooking times given here will need to be adjusted. Similarly, plants harvested at different times of the year or from different places can vary in flavour and intensity, so taste and tweak as you go.

General tips

- Much of what you forage will be similar to garden crops, so eat and cook them in similar ways. But don't be afraid to experiment, as there are some unique and interesting flavours available.
- Some foraged plants, such as *Rumex* species, acquire a metallic flavour when cooked in iron or aluminium vessels.
- In general, the older the leaves are, the tougher or more bitter they will be, so choose fresh growth where you can, preferably before flowering (though the flowers can be useful identifiers).
- If in doubt about the identification of a species, do not use it.
- Avoid anything that might have been sprayed by chemicals or dogs.
- Unless otherwise instructed, make sure your berries are fully ripe, as they can contain toxins if immature.
- To remove stems and leaves from small berries, just freeze in clusters and then use a fork to rake off the unwanted material.
- If a plant is edible, there will be multiple ways you can use it. For instance, edible conifer needles can be shredded and mixed into sweet and savoury dishes; put into pesto and pickles; used for infusing alcohol, salt, sugar or syrup; used in biscuit glazes or ice creams. Keep your mind open!

Flavour properties

When considering how to use your foraging, focus on making the most of the key flavours:

- Acid/sour/citric/lemon flavours are particularly evident in *Rumex*, *Oxalis* and *Berberis* berries.
- Bitterness is provided by brassicas, plantains, older *Rumex* specimens, sow thistles, some Apiaceae, a number of different ferns, turkey tail fungus and the leaves of various different species.
- Heat/mustard/peppery flavours are found in horopito, kawakawa, many brassicas and the mushroom peppery bolete.
- Saltiness predominates in beach spinach, seaweeds, samphire, sea blite and other plants gathered from the shoreline.
- Umami flavours are rarer to find but are strong in mushrooms and seaweeds but also ice plants.
- Sweetness is not just found in fruit but also in Malvaceae species, in porcini and in several other mushrooms.
- Distinct flavours can be found in mints, in aromatic plants such as sweet violets, or as aniseed in fennel and several other Apiaceae plants.

Acid, bitterness or heat will all add zest to blander ingredients like *Amaranthus*, avocado, beans, rice, pulses and potatoes, and they'll cut through rich cream or cheese, fatty meats like lamb, or oily fish or duck. Vinaigrettes and salad dressings were designed to bring oil and acidity to enhance lettuce, and will also do so if your mixed harvestings include bitter leaves. Acid or bitter ingredients will counter the sulphurous flavour in Brussels sprouts and broccoli. Also they will balance sweetness, such as in kūmara, tomato, egg or carrot; or in the case of *Rumex* will go well in sweet dishes paired with apples or gooseberries, in ice cream or in a pancake mix.

Salty foraging, like samphire, will be modified by acid leaves, or will highlight sweetness, or round out the flavour profile of umami dishes; for instance, fungi, samphire and seaweed all work well together.

Sweet pickings, such as elderflowers, will make sweet cordials or desserts, but conversely work well in the acidity of vinegar. Gooseberries can be lifted by mint, plum by fennel, mushroom by mustardy brassicas. In other words, embrace both the mild and the strong flavours.

Preserving and drying

Freezing preserves most of the colour, flavour and nutrients of plants. Clean your foraging, dry, place in bags or containers and freeze. Chopped leaves can be placed in ice-cube trays, topped up with water and frozen. Don't forget to add a date to remind you how long it has been stored.

Many foraged items can be dried and stored in dry, airtight containers placed in a cool, dark cupboard. If you intend to dry your foraging, pick it once the dew has dried, and shake it clean rather than using water. Dehydrators preserve colour, flavour and fragrance. If you don't have a dehydrator, you can use the oven on very low (about 50°C) or a microwave (short blasts at a time, as it is easy to overdo it).

However, another option is to leave your foraging for 5–10 days in a warm, dry position, out of direct sunlight:

- For stemmed plants, make small bunches and hang them upside down; if they have seeds, tie a loose paper bag over the head to collect them.

- For leaves, flowers, roots, bark, fruit and berries, place flat in a single layer on paper towels or muslin laid over a cooling rack. Roots and bark should be cleaned, trimmed and cut into small pieces to help with drying; along with fruit and berries, these will need longer and should be turned regularly. Avoid breaking up any leaves before you use them, to retain their scent.

- Seaweed should be thoroughly washed and patted dry, then hung on a line or laid flat on a rack over tea towels or paper towels in a dry, sunny position (the UV light improves the flavour by breaking down the polyphenols). After a few days it should be dry and crispy and ready to store in airtight containers in a dark cupboard. First, though, you might want to chop or flake it, or reduce it to a powder in a food processor or grinder, depending on how you intend to use it. It will keep for several years. Most seaweeds rehydrate in a matter of minutes once immersed in tepid liquid, but thicker pieces may need up to half an hour.

- Fungi should be wiped clean (avoid water if possible) and sliced or at least halved. Spread on a dry pan in a single layer and place in an oven set to 150°C for about 2 hours, turning once or twice. When completely dry (the pieces will be brittle rather than bendy), seal in a container and they'll keep for several years. Be careful of spores being released, as they can cause allergies. You can dry fungi on a rack in a sunny, dry place over several days, but any humidity will rot them and they are less likely to be properly dried.

- Make sure everything you dehydrate is paper-dry before storing, otherwise mould will grow.

Infusions

INFUSED OILS

Add freshly picked plants to a sterilised clear-glass container and cover with the cooking oil of your choice (no cooking required). Seal the container and place in a sunny position to allow the plant oils to be released. Leave for a couple of weeks, shaking daily. Strain to remove the plant material and bottle, or repeat with fresh plants if you want a stronger flavour. Use for salad dressings, marinades, basting or flavouring vegetables.

- **Savoury oils** chickweed; Chinese quince; fennel; horopito; kawakawa; land cress seed; mint; rimu foliage; sea rocket seed; wild carrot; wild radish
- **Sweet oils** use a nut oil with leaves and flowers of Darwin's barberry; elderflower; flowering currant; gorse flowers; hyssop; kiekie bracts; lavender; lemon verbena; magnolia petals; pineapple weed; rose petals; violet flowers

INFUSED SPIRITS

Choose your spirit base, for example gin or vodka — these are the easiest two to work with, though rum, brandy or whisky can also be used. Steep your chosen foraging in the spirit.

- For highly flavoured plants only a sprig is required.
- For fruit and berries, the more you have in the spirit the stronger the flavour and colour, but also the less time needed for infusing (you don't need to be too exact, but half fruit/berries to half spirit is a rough guide). If you have gone for fruit, prick with a fork before steeping or in the case of hard fruit, such as quince or crab apples, grate.
- For more of a liqueur, also add sugar: about a third to half the amount of fruit or berries; adjust the amount depending on how sweet or bitter the fruit or berries are and how sweet you want the final mix to be. Err on the side of caution, as you can always add more sugar later.

Seal, shake well, then store in a dark place and shake regularly. If you have used highly flavoured leaves or flowers you need wait only 3–5 days. If you are using fruit, berries or milder flowers, then leave for at least a month, preferably three. For a strongly flavoured infusion, leave for longer. In all cases, taste every so often, and when the flavour is to your liking strain out the foraged items and bottle the spirit.

- **Foraged flavours** (you can use a mix and/or add spices): blackberries; chamomile; conifer green cones, needles and resin; damsons; dandelion flowers; Darwin's barberry; elderflower; elderberries; eucalyptus leaves; fennel; flowering currant; gorse flowers; hawthorn berries; heather flowers; horopito leaves; hyssop; gorse flowers; kānuka twigs; kawakawa leaves or fruit; kiekie bracts; lavender flowers and leaves; lemonwood flowers and leaves; loquats; magnolia petals; mānuka twigs; mint; *Prunus* fruit; pineapple weed; red clover; rosehips; staghorn sumac; strawberry tree; violet flowers

INFUSED VINEGARS

Warm but don't boil cider or wine vinegar, and pour over bruised, freshly picked plant material in a sterilised glass container. Seal with an acid-proof lid. Place in a sunny position for a couple of weeks, shaking daily. Repeat with fresh plants if you require a stronger flavour, otherwise bottle (straining off the old plants). Use for salad dressings, marinades and sauces.

Leaves dandelion; comfrey; conifer foliage or needles; herb robert; horopito; hyssop; kawakawa; lemonwood; mint; oxtongue; plantain

Flowers chamomile; Chinese quince; dandelion; elderflower; elderberry; flowering currant; gorse flowers; kiekie bracts; lavender; lemonwood; magnolia; nasturtium; pineapple weed; red clover; rose; scented geraniums; self-heal; violet; wild carrot; wild thyme

Berries/hips/seeds Darwin's barberry; hawthorn; kawakawa; larch catkins; nasturtium; rosehips; taupata

SYRUPS

There are many different recipes and you may need to adjust them to fit different types of foraging. The following is a good place to start for less-juicy ingredients.

First infuse your liquid. Chop your freshly picked foraging and cover with boiling water, then leave to cool and steep for about 12 hours (though taste to check, as you may not need to leave this long). If your foraging is tough, such as rosehips, instead of steeping simmer uncovered for 30–60 minutes, until reduced by about half. Be careful about simmering foliage, as it can become bitter or else lose flavour altogether.

Strain the liquid into a measuring cup, then place in a pan with an equal amount of sugar or honey (or less, depending on how sweet you want it and how tart the ingredients are). Heat and stir until the sweetener has dissolved, then bring to the boil, skimming off and discarding any foam. Bottle and refrigerate.

Use within 2 months, or else freeze in an airtight container. Use to make a sorbet or drizzle over ice cream, cakes and puddings, or add to cocktails or champagne, or dilute to make a cordial.

Syrup flavourings barberries; blackberries; chamomile; conifer needles, resin or unripe cones; Darwin's barberry berries; dandelion flowers; elderflower and elderberries; flowering currant; gorse flowers; hawthorn flowers or berries; Japanese honeysuckle flowers; kiekie bracts; lemonwood leaves and flowers; lilly pilly berries; magnolia petals; mint leaves; nettle leaves; passionfruit; pineapple weed; closed pinecones; pōhutukawa flower stamens; *Prunus* fruit; Queen Anne's lace flower and leaves; red clover; rose petals and hips; rowan berries; scented geranium leaves; staghorn sumac; taupata; sweet viola flowers; wine raspberries

Condiments

CAPERS

You can make capers from two different sources: unopened flower buds (such as from dandelions) or seeds (such as from nasturtiums).

- **Dandelion capers** You want the very new buds still quite small and nestled at the base of the plant within the rosette of leaves. For 1 cup of buds, bring the following to the boil: ⅓ cup vinegar, 2 tablespoons water and ½ teaspoon salt, stirring to dissolve salt. Place the buds in a sterilised jar and cover with the hot liquid. When cool, cover and store in the fridge; use within a couple of months.

- **Nasturtium capers** For 1 cup of nasturtium seeds, steep in a bowl of water with 1 teaspoon salt for 2 days, then drain, rinse and place in a sterilised jar. Simmer 1 cup vinegar with 1 teaspoon sugar; when the sugar is dissolved pour over the seeds, covering completely. Add a bay leaf, garlic clove or herb sprig as desired. When cool, cover and store in the fridge; use within 6 months.

- **Other caper substitutes to experiment with** green elderberries (boil for a few minutes before putting in jar); kahikatea unripe fruit; magnolia flower buds; mallow calyx buds; nipplewort flower buds; onionweed pods; ox-eye daisy flower buds; perennial wall-rocket seedpods; sea rocket seedpods; sow thistle flower buds.

DIPS/RAITA

Dips: Mix plain yoghurt, olive oil, a touch of honey and seasonings with a selection of chopped: kawakawa fruit; onionweed; Vietnamese coriander; walnuts; wood sorrels, etc.

- **Raita** To add to raita (with yoghurt and cucumber—or choko): fennel seeds or leaves; harakeke seeds

HERB BUTTER AND HERB CHEESE

Blitz cleaned, chopped fresh herbs or leaves in a food processor or with a stick-mixer. Alternatively use dried herbs. Mix in slightly softened butter or cream cheese a little at a time so you can reach the flavour you like. Salt or lemon juice can also be added to taste.

Place herb butter in a mould; or roll into a log, refrigerate then cut into slices; or use in garlic bread or to enhance a range of dishes, such as melting over vegetables, baked potatoes, fish or meat.

Store herb butter or cheese sealed in the fridge and eat within several days. Herb butter can be frozen for up to 2 months. You can also add the greens to sour cream.

- **Herbs** calendula petals; chervil; fennel; garlic mustard; geranium leaves; larch needles; mint; narrow-fruited watercress; native celery; onionweed leaves or bulbs; parsley; purslane; salad burnet; sorrel; watercress; wild chervil

QUICK-PICKLING AND BRINING

To pickle, fill a sterilised glass jar with your chosen plant item(s), whole or chopped as desired. Boil apple cider or wine vinegar with salt and pepper and fill the jar to the top. Slightly knock the jar to release any air bubbles. Seal with an acid-proof lid and store the pickle in a dark cupboard. Some items such as magnolia petals will take only a few days, others might need several months before eating. Refrigerate once opened.

Alternatively, you can ferment your foraging in brine (1 tablespoon salt to 1 cup water to roughly 1 cup of foraging). Mix salt into water, pour over foraging in a container, cover with cheesecloth and leave for a couple of weeks. When it stops bubbling, put on a lid and store in the fridge.

Suitable for pickling black tree fern koru; barberries; choko; conifer catkins; unripe harakeke seedpods; kawakawa leaves and fruit; magnolia buds and petals; mushrooms; Neptune's necklace; onionweed stalks or bulbs; pikopiko; purslane; quince fruit; samphire; sea blite; sea rocket; seaweed bladders, fronds, stems and tubes; South African ice plant fruit; thistle flower buds; walnuts; wild radish.

PEPPER/SALT/SPICE SPRINKLES

Nothing can be easier than dehydrating your foraging (see above), then blitzing it in a coffee grinder to a fine powder. Store in a pepper shaker, either alone or mixed with salt or other ingredients to add flavour to your cooking.

The mushrooms in this book will all provide an umami flavour, while the dried seaweeds will give a taste of salty iodine, though for a more nutty flavour mix in ground and toasted wild grass seed, linseed or mallow seeds.

Horopito, kawakawa, and watercress leaves will produce a bit of zest or fire, as will peppery bolete or the seeds of purslane, sea rocket, staghorn sumac and watercress.

Also try conifer needles for an aromatic resinous flavour or kānuka and mānuka seeds, or for something more, herby: bay leaves, geranium seeds, parsley flowers and seeds, wild carrot.

CHUTNEY

Slice fruit, removing stalks, stones and in some cases the seeds (this will also need to be done if rosehips are being included as the berry component). Chop onions and mix with fruit, berries, peppercorns and salt (see quantities below). Leave covered overnight.

Bring sugar and vinegar to the boil, add fruit mixture and simmer until thick. Bottle in sterilised jars and leave for at least 2 months before consuming.

Quantities and suitable foraging:

- 1kg fruit (replacing or combining with green tomatoes or apples): blackberries; black nightshade; Cape gooseberries; Chilean ice plant figs; crab apples; elderberries; figs; flowering currant berries; gooseberries; lilly pilly berries; medlars; poroporo; *Prunus* spp.; quince; wine raspberries
- 250g dried (or fresh) berries (replacing sultanas and raisins): barberries; hawthorn; rosehips, rowan

- 2 small onions (onion weed could replace some of this)
- 10g (replacing peppercorns or spices) seeds from: hedge mustard; honesty; kōhia; nasturtium; shepherd's purse; watercress; water pepper (ground peppery bolete could be used instead, or conifer needles)
- 15g salt
- 200g sugar (adjust depending on how sweet the fruit component is)
- 2 cups vinegar

SAUCE/KETCHUP/DRESSINGS

For sauce or ketchup, bring the following to the boil, then simmer for 30 minutes:

500g foraged fruit or berries
300ml apple cider vinegar
300ml water
½ teaspoon salt
pepper
a pinch of any spices of your choice

Press through a sieve. Discard pulp and seeds, then return the rest to the pan with 175g sugar. Boil, stirring, for 10 minutes. Stored in sterilised jars, it should keep unopened for 6 months. Refrigerate when opened.

- **Fruit or berries**: try crab apples; blackberries; elderberries; hawthorn berries; lilly pilly berries; loquat; medlar; quince; rosehips
- For a **hot version**, try horopito leaves or onionweed

For a quick, strongly flavoured salad dressing, try blending oil, vinegar, a touch of sugar, and salt and pepper with your selection of the following:

- horopito; kawakawa leaves; nasturtium leaves; mint leaves; onionweed; salad burnet; sea rocket; wood sorrels

For a green sauce or salsa verde, chop and mix a couple of cloves of garlic, with a handful selected from your foraged capers and pickles, together with some handfuls of foraged leaves. Add mustard, vinegar and stir in olive oil and seasoning to taste. Foraged leaves can include:

- chickweed; dandelion; docks; ferns; hedge mustard; Italian parsley; mint leaves; nasturtium; native celery; native ice plant; onionweed; plantains; salad burnet; sorrels; stinging nettles; tradescantia; Vietnamese coriander; watercress; water pepper; wild chervil

For mint sauce, dissolve a tablespoon of sugar in a couple of tablespoons of boiling water, finely chop a good handful of mint leaves and add to the sugar water along with a tablespoon of wine vinegar. Mix and leave for the flavours to combine for an hour, then consume.

Savoury dishes

FOR THE POT

For cooked vegetables, adding to casseroles, using instead of spinach or Asian greens, etc.:

- alexanders; amaranths; barberries; black tree fern koru; cabbage tree; campions; chickweed; choko; cleavers; green conifer cones and sprigs; dandelion; deadnettle; docks; various ferns; ground ivy; hedge mustard; hosta; Japanese honeysuckle; kahikatea; linden tree; mallows; mushrooms; native celery; orache; plantains; puha; samphire; self heal; sorrels; seaweeds; stinging nettle; sweet violet; thistle stem and leaves; tradescantia; twin cress; watercress; wild asparagus; wild fennel; wild parsnip

Most leaves can be steamed as they will require very little cooking, though boiling in water (which should be discarded) will reduce bitterness and in some cases remove toxicity. Tossed in lemon juice and olive oil, they make an excellent side dish.

Most seaweeds and fungi can also be eaten with minimal input.

CHIPS

Just as you can make kale chips, so you can make chips from many foraged leaves or thinly sliced roots. Just douse the leaves or roots with olive oil and your favourite spices, spread them individually on baking paper and bake until crisp (anything from 3 to 15 minutes — watch to avoid burning) at 150°C. Turn to ensure even drying. Add salt and spices as desired and eat!

- **Try** plantain; most seaweeds; wild parsnips

PESTO

Traditional ingredients are: pinenuts, garlic, basil, Parmesan, olive oil, salt. Proportions can be to taste, but approximately:

1 part nuts
1 part garlic
2 parts oil (add more to make runnier if required)
2 parts cheese
6–8 parts basil
plus a generous pinch of salt

Whizz together in a food processor or with a stick-mixer. Add to pasta as a sauce, to pizza as a base, to salads as a dressing, to bread or crackers as a spread. Once made it can be placed in ice-cube trays and frozen for use later.

Substitutes:

- **Instead of pine nuts** any nuts (e.g. walnuts, Brazil nuts or almonds); buckwheat, sunflower or pumpkin seeds; Monterey pine nuts

- **Instead of garlic** bulbs and/or leaves of onionweed; garlic mustard

- **Instead of oil** avocado (add with a little water to give you the consistency you require)

- **Instead of Parmesan** any cheese, nutritional yeast or miso
- **Instead of basil** amaranth; choko; bird's eye speedwell; bitter cress; black shield fern; water celery; chickweed; cleavers; dandelion leaves; dock leaves; hedge mustard; linden leaves (young); miner's lettuce; nasturtium leaves; parsley; plantain; puha; purslane; wild rocket; self heal; sow thistle; storksbill; sorrel; spearmint; New Zealand/beach spinach; stinging nettle; Vietnamese coriander; wall lettuce; watercress; wavy bittercress

Note: lemon juice and/or cider vinegar and/or honey can be added if leaves are too bitter, or use a variety of plants, or increase the quantity of nuts.

SALADS

There's nothing easier than pulling together a mix of raw foraging. Choose young leaves, as they will be less bitter and have a more tender texture. But don't be put off by bitter leaves as they can add an extra dimension when mixed with leaves with other flavours, such as spicy nasturtium, salty beach spinach or fresh mints, and then balanced out by blander leaves, such as miner's lettuce and tradescantia. There are so many species in this book that will provide leaves to go into your mix, and don't forget there are also edible flowers to add both brightness and flavour to your mixture of leaves.

For a seaweed salad, you can simply dress your cleaned fresh seaweed with sesame seeds and an oil — either sesame oil or olive oil are ideal. Or for a variation, dress the leaves with a mixture of sesame oil, onionweed and garlic.

SOUPS

(Serves 4) Heat 50g butter or oil, add 1 finely chopped onion and fry; when translucent add 2 diced potatoes and 4 cups of your choice of leaves (either all one species or a mixture; discard any woody stalks). Cook for a couple of minutes. Other chopped vegetables could also be added at this stage, such as carrots, celery, etc.

Add 1 teaspoon vegetable stock and bring to the boil. Simmer until the potatoes are soft, then blitz with a stick-mixer or in a food processor or liquidiser. Season to taste with salt, pepper and grated nutmeg and garnish with grated cheese, yoghurt or croutons.

As with salads, there are numerous plants whose leaves make excellent additions to soups, some — such as nettles — being classics. Don't forget that mallow leaves and lesser burdock roots contain useful thickeners. For the chopped vegetables, you can use choko, wild carrot, parsnip or other roots you have foraged. You can also use a variety of seaweeds, fungi or berries.

Desserts

DECORATIVE EDIBLE FLOWERS

Drop rinsed flowers into ice-cube trays to create attractive ice cubes for your drinks.

Or crystallise to decorate slices and cakes: dip them into the lightly whisked white of an egg, remove excess so you have only a thin coating, dowse with icing sugar, and leave to dry in a cool place on greaseproof paper for at least 24 hours. Use within a few days.

- **Flowers** cabbage tree; camilia; Chinese quince; gorse; hawthorn; ivy-leaved toadflax; magnolia; oxalis; pansies; rose petals (remove bitter white base); sweet violets; wild pansy

- Angelica stems, wild fennel leaves and barberries can also be candied

SAVOURY OR SWEET JELLY

For fruit jelly: Simmer together 2kg chopped crab apples or other foraged fruit, 275ml wine vinegar, 800ml water (or less if the fruit is particularly juicy), and a handful of selected fresh herbs. When fruit is soft, gently strain off the juice through a muslin overnight (do not squeeze). Measure the juice and place in a saucepan with 60g sugar per 100ml of juice. Boil until it foams up and setting point is reached, then cool for 10 minutes, fill sterilised jars, seal and store in a dark cupboard.

For herb jelly: Roughly chop 450g of your chosen herb or mixture of herbs, add enough water to cover generously, and simmer for 1 hour (checking that it does not run dry). Add the juice from 2 lemons and simmer for another 10 minutes. Remove the herb by straining through muslin. For each cup of liquid, add a cup of sugar and simmer until it reaches setting point. Transfer to jars as above.

For petal jelly: Collect 4 cups or 1000ml petals, rinse, then cover with 4 cups boiling water and steep for 10 minutes. Add 2 tablespoons lemon juice, stir and strain into a saucepan, removing the petals. Bring to the boil, stir in 50g pectin, and allow to boil for 1 minute before stirring in 4 cups sugar. Boil for another minute, then ladle into sterilised jars. Seal and store as above.

- **Fruit** barberries; blackcurrant; Chinese quince; elderberries; flowering currant; hawthorn berries; lilly pilly berries; loquat; medlar; quince; rowan berries; South African ice plant; staghorn sumac

- **Herbs** apple mint; chamomile; elderflower; hawthorn berries; lemon balm; mint; parsley; Queen Anne's lace flower

- **Petals** (remove all other parts of flower): bee balm; clover; dandelion, elderflower; flowering currant; hibiscus; honeysuckle; kiekie bracts; lavender; milkweed; nasturtium; red clover; rose; self-heal; violets; willow herb

Jams

FRUIT OR BERRY JAM

Not all fruits and berries have high pectin levels, so it's worth including apple, crab apple, plums or grated lemon peel with your foraging to ensure your jam sets.

Recipes vary, but as a guide use the same weight of fruit/berries as sugar and adjust to your taste. Use about 150ml water per 1.5kg of sugar, unless you're using juicy fruit or berries, in which case little or no water is required.

Bring chopped fruit or berries to a simmer with the water (if using). Stir in sugar, and once dissolved boil rapidly for anything from 10 minutes to an hour, depending on how soft the fruit/berries are. Once setting point has been reached, cool for about 10 minutes and bottle in sterilised jars.

Fruit barberries; blackberries; blackcurrants; black nightshade; bush lawyer berries; Cape gooseberries; Chinese quince; elderberries; flowering currant; fruit-salad plant; gooseberries; hawthorn berries; Himalayan honeysuckle; kawkawa berries; loquat; lilly pilly berries; mock strawberries; passionfruit; poroporo; prickly pear; *Prunus* spp.; quince; rowan berries; South African ice plant; snowberries; snow tōtara; strawberry tree; tree fuchsia; wine raspberries

PETAL JAM

Add 2 cups petals to 375ml boiling water and simmer uncovered for 10 minutes.

Stir in 2 cups caster sugar and 3 tablespoons lemon juice and simmer for 10 minutes.

Sprinkle in 1 teaspoon pectin and simmer for another 20 minutes. As an option you can add 1 teaspoon rosewater, orange blossom water or hibiscus flower water, or a few drops of lavender oil.

Cool for 10 minutes, then fill sterilised jars.

Petals black locust tree; elderflower; hibiscus; lavender; rose (with bitter white base removed); violet

Baking

ADDITIONS TO BREAD DOUGH, CRACKERS AND SEED MIXES

Add to your chosen recipe:

- chopped conifer needles; dock leaves; fennel seeds; dried and ground hawthorn berries; linseed; onionweed leaves; pine pollen; plantain seeds; purslane seeds; dried and ground rosehips; seaweed; sorrel seeds; walnuts; wild carrot; wild fennel; wild grass seeds

FLOUR

For a recipe that needs to rise, such as leavened bread, you ideally need a flour with gluten in it, such as wheat. However, flour without gluten can be mixed in with wheat or else used for flatbread, porridge, crackers, biscuits, etc.

It's best to roast your grains in the oven on a low heat for about 5 minutes to dry them thoroughly and bring out their nuttier flavours, then process in a coffee grinder. Husks are fine to include, as they add fibre.

Flour can be made from a variety of plants (and as some can be bitter or found only in small quantities, mixing your grains is often a good idea):

- acorn (see note); alder catkins; amaranth; bamboo seeds; birch catkins; bulrush pollen; burdock root; chestnuts; clover seeds; cottonwood catkins; crabgrass seeds; dock seeds; dried and ground hawthorn berries; karaka kernels (see note below); lambs quarter; larch cambium (inner bark); lichen; linden cambium; linseed; maple cambium; orache; pine bark; pine pollen and catkins; plantain seeds; prickly pear seeds; purslane seeds; raupō pollen; sea rocket root; sheep's sorrel root and seeds; silver birch cambium; sweet chestnuts; Sydney golden wattle seeds; tawa kernel; wattle seeds; wild grass seeds

- You can also add pikopiko to doughs

Note: Acorns must be soaked in cold water for several days to leach the tannins. Dry them first for a few weeks, then crack the shell and purée the kernel with water and keep cool (refrigerate if you have room), replacing the water every day for a week. It is easier to change the water if you place the puréed kernel in a tea towel inside the bowl of water and squeeze out the water for each change. Soak them until the water becomes clear. Then dehydrate and grind.

Karaka kernels must be prepared following strict guidelines to remove toxicity. Although it is wise to follow someone experienced in the process, here is an indication of what is required. Select only ripe berries, simmer covered in water for up to 8 hours, topping up with water as needed. Cool and remove the flesh, then soak the kernels in cold water for 5 days, changing the water each day, then dry, roast, remove the outer shell and grind the inner flesh into flour.

Drinks

CORDIAL AND KOMBUCHA

See the syrup suggestions above. For cordial, dilute to taste with still or soda water or a mixture of lemon juice, water, and citric acid. For kombucha, follow your usual recipe, then add your own combination of the foraging, mashed or juiced, to the second fermentation.

COFFEE SUBSTITUTES

For root coffees: Select large plants from uncontaminated soil. Dig up the roots and wash them thoroughly. Slice the roots and roast them dry for 2–4 hours in a low oven at about 100°C until you can snap them. Store in a sealed jar, and when you are ready to brew your coffee take about 2 dessertspoons and grind in a coffee grinder. Either use in a coffee plunger or for a stronger brew place in a saucepan with 2 cups of water and simmer for 2 minutes. Strain into your mug and enjoy with or without your usual additives.

You can mix roasted roots from different species, and you can also try seeds instead (the roasting time will be a lot shorter) and also add in other ingredients such as cardamon, chicory or cinnamon.

- **Roots** catsear; dandelion; lesser burdock; wild carrot
- **Seeds** cleavers; coprosma; dock; hawthorn; hibiscus; loquat; sow thistle; variegated thistle; wild asparagus
- **Nuts** acorns; sweet chestnuts

SMOOTHIES

Blend a mixture of green leaves together with anything from 1 to 4 cups of your chosen liquid, depending on how thick you like it (and how much you are making). Once the leaves are well chopped, blend in the fruit (the ratio depends on your preference, but on average ¾ fruit to ¼ leaves). When smooth, serve and eat immediately.

- **Green leaves** amaranth; bird's-eye speedwell; bittercress; chickweed; cleavers; comfrey; red clover flowers and leaves; dandelion; herb robert; native celery; New Zealand spinach; plantain; purslane; puha; sorrel; salad burnet; sow thistle; stinging nettles; tradescantia; violet leaves; water celery; wild fennel; wild silverbeet
- **Fruit** banana; barberries; frozen blackberries (which will chill the smoothie); loquat; *Prunus* spp.; wine raspberries
- **Other foraging** black tree fern koru; linseed; pine pollen; raupō pollen
- **Liquids** water; yoghurt; nut milk; milk; ice cream

TEAS

Teas can be made from fresh or dried leaves, flowers and/or roots. Simply boil in water for anything from 2 to 30 minutes until the desired strength of flavour is achieved. Strain and drink hot.

Tea can be turned into a cold drink if cooled, mixed with ice, sparkling water and a sweetener such as honey or berry syrup, and extra flavour such as lemon juice or ginger. To become more of a cordial, steep your tea with its foraging for 24 hours and add ½ cup sweetener (e.g. sugar or honey) to 4 cups tea.

- **Leaves** bergamot; bird's-eye speedwell; blackberry (though some people will react to this); blackcurrant; conifer needles; creeping wood sorrel; dead nettle; eucalyptus; gooseberry; hawthorn; herb robert; kānuka; kawakawa; lemon balm; mallow; mānuka; mint; Portuguese angelica; nipplewort; plantain; salad burnet; shepherd's purse; silver birch; sorrel; stinging nettle; wild carrot; wild fennel; wild strawberry; yarrow

- **Flowers**bird's-eye speedwell; black locust tree; elderflower; hibiscus; linden tree; mallow; Queen Anne's lace flower; red clover; rose petals; sweet violet; yarrow

- **Berries** elderberry; rosehip (these need to be blitzed in a food processor first, then strained through a fine mesh several times to remove the fine hairs on the hips, which can be an irritant)

- **Other** mallow seeds; reindeer moss; turkey tail fungus.

Index

Common Names and Māori Names

Botanical Names

Acknowledgements

I would like to thank the following people who made *Foraging New Zealand* possible:

To my parents, Ian and Elizabeth Langlands, for taking me into the outdoors as a child, and for our trips into the Canterbury high country as a teenager, foraging for wild spearmint, gooseberries, blackberries and apples, and encouraging my love of birdwatching. To Martin Tickner, who introduced me to the sheer diversity of wild foods available in New Zealand, and specifically on my foraging homeground of Banks Peninsula. It was truly through you, Martin, that my wild food journey was launched, and my confidence to forage wild fungi was developed. To Chef, Vaughan Mabee, for the opportunity to forage commercially for Amisfield restaurant in Central Otago and to supply foraged foods to the restaurant scene in New Zealand. This opportunity to work with chefs truly opened my eyes to the full scope of using wild foods. To my partner in crime, Jackie Phillips — I'm grateful for the passion we share for foraging and travelling, exploring New Zealand's landscape, and conducting foraging workshops. Jackie's creativity in using wild food has nourished me not only intellectually but, fuelling many trips into the outdoors! To my editor, Harriet Allan, for her passion in instigating this project and energy in assisting with its completion, as well as the team at Penguin, especially Rachel Eadie and Faith Wilson for managing the production of this book, and Carla Sy, for her design. To Jon Sullivan, for his expert input and review of my manuscript, and for his passion in recording New Zealand's biodiversity with the iNaturalist.NZ database. To Monica Wilde, whose instincts and insights into foraging food safety are unmatched. Finally, to the many people who have attended my foraging workshops and webinars over the years. You have enabled me to make a living as a professional forager and to focus on this exciting field in which there is always something new to discover. I hope you see this book as a gateway to exploring the many thousands of edible species that we have in New Zealand.

— Peter Langlands

About the Author

With a lifelong interest in the outdoors, Peter spends his time in a wide variety of environments that present a diversity of foraged foods, from alpine to coastal locations. After studying natural sciences at Canterbury University and post-graduate papers in zoology, foraging turned him from a zoologist to an accidental botanist. Peter now works as one of Aotearoa's only licensed professional foragers, sourcing wild produce for chefs around the country, including the innovative Amisfield, running foraging workshops and sharing his love of the subject through his social media pages and website @wildcuisinenz. Peter has spent a lifetime compiling Aotearoa's largest database of wild foraged species and works with international foragers on expanding the spectrum and use of wild foods. Peter continues to discover and explore the abundance of food Aotearoa has to offer, with his partner from their Ōtautahi Christchurch base.

Additional Resources

BOOKS

A Field Guide to the Native Edible Plants of New Zealand, Andrew Crowe (2004)

An Illustrated Guide to Common Weeds of New Zealand, Ian Popay, Paul Champion, Trevor James (2010)

Fungi of Aotearoa: A Curious Forager's Field Guide, Liv Sisson (2023)

Hiakai: Modern Māori Cuisine, Monique Fiso (2020)

New Zealand's Native Trees, John Dawson, Rob Lucas, Jane Connor, Barry Sneddon (2019)

New Zealand Seaweeds: An Illustrated Guide, Dr Wendy Nelson (2020)

Simply Living: A Gatherer's Guide to New Zealand's Fields, Forests and Shores, Gwen Skinner (1981)

The Forager's Treasury: The essential guide to finding and using wild plants in Aotearoa, Johanna Knox (2020)

The Meaning of Trees, Robert Vennell (2019)

The Wilderness Cure, Mo Wilde (2022)

Wild Mushrooming: A Guide for Foragers, Alison Pouliot (2021)

RESEARCH PAPERS

'Plants in New Zealand Poisonous for Children', Manaaki Whenua – Landcare Research (2002)

WEBSITES

iNaturalist NZ — inaturalist.nz

Māori Plant Database — rauropiwhakaoranga.landcareresearch.co.nz/

National Poisons Centre — poisons.co.nz

New Zealand Weeds Database — massey.ac.nz/about/colleges-schools-and-institutes/college-of-sciences/our-research/themes-and-research-strengths/plant-science-research/new-zealand-weeds-database/

Plants for a Future — pfaf.org

ONLINE COMMUNITIES

Foraging New Zealand — facebook.com/groups/363115978004948

Mushroom Hunting New Zealand — facebook.com/groups/mushroom.hunting.nz

Seaweed Foraging New Zealand — facebook.com/groups/324733514387077

Wellington Foraging Community — facebook.com/groups/559875947538402

Wild Capture — Wild Foods and Foraging NZ — facebook.com/forageNZ

Image Credits

All photographs © Peter Langlands, except:

Alamy: 21, 135, 137

Evelyn Simak: Hedge Woundwort Photo, CC BY-SA 2.0, 171

Flickr: Björn S..., flickr.com/photos/40948266@N04/37793987452, 439 (below), Docentjoyce, flickr.com/photos/docentjoyce/50010998777, 359 Katja Schulz, flickr.com/photos/treegrow/45270077944, 367

Harriet Allan: 129 (above left and right), 360, 361

iStock: 39, 57 (above)8, 81 (below), 102, 103, 105, 130, 133, 136, 159, 163, 188, 189, 205, 213, 227, 231, 239, 243, 245, 247, 262, 295, 296, 313, 323, 327, 335, 348, 353, 369, 400, 479 (below left)

Jessica Davies: 463

Jon Sullivan: 23, 93, 113, 127, 195, 204, 206, 207, 253, 257, 261 (below), 287 (top right, bottom), 297, 322, 373 (below), 447, 478 (below left), 479 (top right), 480 (middle, below), 481

Shutterstock: 16, 19, 29 (insert) 31, 35, 54, 57 (below), 91 (full page), 95, 101, 119, 150, 156, 225, 239 (inset), 255 (top left), 256, 289, 303 (below right), 341 (below), 347, 373 (above), 375, 385, 389, 449, 453 (above)

Wikimedia Commons: CC BY-SA 4.0, creativecommons.org/licenses/by-sa/4.0, Dermorgendanach 121 (insert), Krzysztof Ziarnek, Kenrai 104, 50, 51, 383, Salicyna 227 (insert), Stefan Iefnaer 104, Tomas Sobek 311.

Wikimedia Commons: CC BY 2.0 DEED creativecommons.org/licenses/by/2.0/deed.en, Harry Rose 231 (below right)

Wikimedia Commons: CC BY-SA 3.0 DEED, creativecommons.org/licenses/by-sa/3.0/deed.en Barq 194, Alvesgaspar, 332

Purslane

The information contained in this book is of a general nature only, and foraging for ingredients should be treated with caution. If you do wish to make use of any information in this book, you should first consider the appropriateness and safety to do so, including the possible need to consult an expert.

PENGUIN

UK | USA | Canada | Ireland | AustraliaIndia | New Zealand | South Africa | China

Penguin is an imprint of the Penguin Random House group of companies, whose addresses can be found at global.penguinrandomhouse.com

Penguin Random House New Zealand

First published by Penguin Random House New Zealand, 2024

10 9 8 7 6 5 4 3

Design by Carla Sy © Penguin Random House New Zealand
Cover image by Pixel Stories, Stocksy
Prepress by Soar Communications Group
Printed and bound in China by Toppan Leefung Printing Limited

A catalogue record for this book is available from the National Library of New Zealand.

ISBN 978-0-14377-589-8

penguin.co.nz

MIX
Paper | Supporting responsible forestry
FSC® C104723
FSC www.fsc.org